基础设施之光
INFRASTRUCTURE

建筑立场系列丛书

[英]扎哈·哈迪德建筑师事务所 等 | 编
司炳月 高松 | 译

大连理工大学出版社

基础设施之光

建筑立场系列丛书 No.95

- 004　从"路毙"现象说到基础设施带来的暴力与安宁 _ Richard Ingersoll
- 016　阿尔托大学地铁站 _ ALA Architects + Esa Piironen Architects
- 026　先锋村站 _ aLL Design
- 038　普林斯顿车站大厅和商店 _ Rick Joy Architects
- 050　拉赫蒂枢纽站 _ JKMM Architects
- 062　Nørreport车站 _ Gottlieb Paludan Architects + COBE
- 076　洛里昂布列塔尼火车南站 _ AREP
- 088　那不勒斯阿夫拉戈拉高铁站 _ Zaha Hadid Architects
- 104　香港西九龙站 _ Andrew Bromberg at Aedas

- 122　城市基础设施走向光明的未来 _ Phil Roberts
- 132　盎格鲁湖中转站和广场 _ Brooks + Scarpa
- 148　停车楼和屋顶 _ JAJA Architects
- 164　卡特韦克海滨地下停车场 _ Royal HaskoningDHV + OKRA Landscape Architects
- 178　里斯本游轮码头 _ Carrilho da Graça Arquitectos
- 194　波尔图游轮码头 _ Luís Pedro Silva, Arquitecto, Lda.
- 214　西港2号码头大楼 _ PES-Architects
- 228　克罗顿滤水厂 _ Grimshaw Architects

- 238　建筑师索引

Infrastructure

C3 No.96 Infrastructure

004   Road Kill. On the Violence and Pacification of Infrastructures _ Richard Ingersoll

016   Aalto University Metro Station _ ALA Architects + Esa Piironen Architects

026   Pioneer Village Station _ aLL Design

038   Princeton Transit Hall and Market _ Rick Joy Architects

050   Lahti Travel Center _ JKMM Architects

062   Nørreport Station _ Gottlieb Paludan Architects + COBE

076   Lorient Bretagne South Railway Station _ AREP

088   Napoli Afragola High Speed Train Station _ Zaha Hadid Architects

104   Hong Kong West Kowloon Station _ Andrew Bromberg at Aedas

122   Toward a Bright Future of Urban Infrastructure _ Phil Roberts

132   Angle Lake Transit Station and Plaza _ Brooks + Scarpa

148   Parking Houses + Konditaget Lüders _ JAJA Architects

164   Underground Parking in Katwijk aan Zee _ Royal HaskoningDHV + OKRA Landscape Architects

178   Lisbon Cruise Terminal _ Carrilho da Graça Arquitectos

194   Porto Cruise Terminal _ Luís Pedro Silva, Arquitecto, Lda.

214   West Terminal 2 _ PES-Architects

228   Croton Water Filtration Plant _ Grimshaw Architects

238   Index

基础设施之光

从"路毙"现象说到基础设施带来的暴力与安宁

从"路毙"现象说到基础设施带来的暴力与安宁
城市基础设施走向光明的未来

Road Kill. On the Violence and Pacification of Infrastructures / Richard Ingersoll
Toward a Bright Future of Urban Infrastructure / Phil Roberts

RUCTURE
acification of Infrastructures

从"路毙"现象说到基础设施带来的暴力与安宁
Road Kill. On the Violence and Pacification of Infrastructures

Richard Ingersoll

自人类最初决定以定居的模式生活后，就建立了各种形式的基础设施，有古代哈拉派（今位于巴基斯坦地区）文明所创造的惊人的水闸遗址，有中国汉代于首都长安修筑的带有排水系统的大道，还有位于西班牙塞哥维亚雄伟壮观的古罗马水道桥（图1）。尽管如此，人们普遍使用"基础设施"这个名词还是19世纪中期之后的事情。与此同时，随着这个术语的使用，人类近乎粗暴地开始建设铁路、下水管道和高速公路。19世纪60年代，城市设计大师巴伦·豪斯曼为了建造林荫大道和拱形下水道，像外科医生给病人做手术一样，对巴黎"开了刀"，"切除了"这座城市体内大量的中世纪废墟。而在同时代的伦敦也有好几个区域被"肢解"，来支持世界上第一个地下铁路系统的建设工作。建设者只顾着提高效率，加快速度来建设更多干净整洁的设施，使交通更顺畅，使供水系统和排污系统更发达，却给当地的居民和他们的生活用地造成了无情的破坏。以美国铺设高速公路为例，修筑1.6km长的高速公路需要夷平约16ha的土地，施工地区经常是没有被开垦的荒地和贫民窟。"路毙"（Road Kill）一词原本是用来委婉地表示被马路上疾驰的汽车撞死后的动物尸体，但考虑到美国在过去80年内，平均每天有100个人死于车祸的这个事实，这个词似乎也可以用到那些人身上，因为他们也是基础建设过程中的受害者。也许听起来有点对社会冷嘲热讽的意思，但它对基础设施在建设和使用的过程中反复出现的暴力现象可谓形容得恰如其分。

1909年，菲利波·马里内蒂发表了《未来主义宣言》，宣言中他指出了工业化世界中存在着一种心理，那就是对一切的淡漠和潜意识里对死亡的追求。他提到，有一次开车时，因试图避开两个过往的骑行者而导致一场车祸。他感情激昂地描述了这段经历，"我

While various forms of infrastructure have existed since the earliest human settlements, one thinks of the impressive sluices in ancient Harappa, the broad avenues lined with drainage ditches in Han Dynasty Chang-An, or the majestic Roman aqueduct in Segovia[figure 1], the word did not come into common usage until the mid 19th century, coincident with the increasingly violent practices of building railroads, sewers, and highways. During the 1860s Baron Haussmann, using surgical metaphors, "disemboweled" the dense Medieval fabric of Paris to make his broad planted boulevards and vaulted sewers, while sections of London were blown to bits to produce the world's first underground rail system. The imperative for efficient, fast, and hygienic services, for the movement of vehicles, water, and waste, resulted in ruthless damage to the land and the urban dwellers. To build 1.6km of the American freeway, for instance, required scraping away about 16ha, often plowing through virgin land or poor neighborhoods. "Road kill" was coined as a euphemistic expression referring to the carcasses of helpless animals inadvertently slaughtered by speeding cars. To use the same moniker for the social and natural life that has been endangered through building infrastructures – consider for instance the average of 100 human deaths per day by automobile accident during the past 80 years in the US – may sound slightly cynical, yet it perfectly conveys the cycle of violence that goes into the production and use of infrastructure.

Filippo Marinetti in his *Futurist Manifesto* of 1909 captured both the endemic carelessness and the inherent death wish present in the new world of industrial culture. After describing in breathless ecstasy his involvement in an automobile accident, caused by trying to avoid hitting two humble cyclists, he proclaimed: "We declare that the

图1 古罗马水道桥，西班牙塞哥维亚
fig.1 Roman aqueduct in Segovia, Spain

们发现了一种新的魅力，它让我们的世界更为辉煌，那就是速度的魅力。比如，汽车飞驰时，引擎盖下的管路像数条盘踞的巨蟒喷着热气……而咆哮的马达犹如喷火的机关枪一样，它比位于希腊萨摩色雷斯岛的胜利女神更加美丽。"虽然他对技术的酷爱在当时人们的眼里显得惊世骇俗，但他的思想主宰了后来的大半个20世纪的西方世界。现代建筑大师，勒·柯布西耶就是《未来主义宣言》的读者和一个积极响应者，他在1923年自信地说："建设高速运转的城市就是在建设成功的城市"。他还说他梦想有一天能够成功打造出供300万人生活的现代都市，这个都市的公路系统中没有街道，因为街道已被16条车道的高速公路取代了。

诚然，勒·柯布西耶提出了如上设想，但如果我们就因此把责任都推给他，认为正是他把此后的城市变成如此形态的始作俑者，那也未免把他作为建筑师的职能想得太夸张了。其实，城市发展方案大多是由汽车行业为满足其需求公然提出的。比如那些默默无闻的实用主义工程师们，他们并不知道勒·柯布西耶是谁，他们的名字也不会被人们铭记在心，但他们依然近乎失控地响应工业文化的指令，追求速度。不过，也有些人例外，他们不那么激进，比如，弗里兹·托特。在20世纪30年代，他负责执行希特勒的"道路计划"。他们剥削集中营俘虏的劳动力，铺设连接柏林和纽伦堡之间的公路，幸运的是沿线的重要景观没有遭到暴力破坏而得以保留。在同一时期的美国，工程师吉尔摩·大卫·克拉克有几条公路也设计得十分出色。当时罗斯福政府为救助失业者而实施了产业振兴政策，1938年由吉尔摩·大卫·克拉克设计的梅里特公园路就是这个政策的一环，它位于康涅狄格州的郊外地区，是当地一道优美的风景线。

splendor of the world has been enriched by a new beauty: the beauty of speed. A racing automobile with its bonnet adorned with great tubes like serpents with explosive breath ... a roaring motor car which seems to run on machine-gun fire, is more beautiful than the Victory of Samothrace." His technophilia, while shocking for the times, became the reigning mentality for most of Western civilization during the twentieth century, echoed by Le Corbusier (one of the readers of the *Futurist Manifesto*), who proudly stated in 1923: "the city made for speed, is the city made for success." He promoted a dream of the "contemporary city for 3 million people" structured around a 16-lane freeway but having no streets.

To blame Le Corbusier's vision as the determinant of later urban form perhaps gives too much credit to the architect, since most solutions evolved as the naked expression of the automobile industry's demands. Pragmatic engineers, who had never heard of Le Corbusier and whose names will never be remembered, followed industrial culture's mandate for speed, almost unbridled. There were some gentle exceptions, such as the banked curving roads introduced by Hitler's highway organizer Fritz Todt in the 1930s, which despite their exploitation of slave labor from the concentration camps had the virtue of conserving significant landscapes on the route between Berlin and Nuremburg. During the same years in the US Gilmore D. Clark designed several exquisite highways, such as the Merritt Parkway (1938), which laces itself gracefully through Connecticut's rural landscapes, built with the participation of Roosevelt's Works Progress Administration for the unemployed.

The US had an obvious advantage over the rest of the world as the first fully automobilized nation – already

图2 自由大道公园，美国西雅图，1972—1976年
fig.2 Freeway Park, Seattle, USA, 1972-1976

美国早在20世纪20年代人均汽车拥有率就达到了五分之一，之后更是与其他国家拉开显著差距，成为第一个汽车全面普及的国家。美国汽车之所以能够那么早普及，得益于政府的一项决策——1956年美国政府启动数十亿美元的预算，开始建设长达66 000km的公路体系。期间，一些开发商以投机利益怂恿政府规划部门做出了一些权宜之计，把很多废物垃圾就扔到了像波士顿、纽瓦克这样的中心城市。一些工程以"城市改造"为名，对开发之外的事情毫不关心，为了限制这些工程模式，劳伦斯·哈普林等一批景观设计师提供了初步方案。他们从1968年出版《高速公路和城市》开始，就一直在思考如何打造一条高速公路，使其与大自然融为一体，同时也符合中心城市的环境要求。哈普林致力于减缓公路给人们带来的消极影响，的确称得上是这批公路建设者中最敬业的一位。西雅图自由大道公园（图2）于1972年动工并于1976年竣工，施工期间，哈普林在环绕市中心的八车道公路上像编织织物一样编织了面积约2ha的混凝土。他在公园种植了几百棵高大树木，其中有些是红豆杉。这些树木被集约式地栽种在混凝土盆里，并在周围设置了喷泉，这既吸收了二氧化碳，又减少了车辆噪声。

虽说现代基础设施的建设和使用的确产生了负面的影响，也的确引发了一些骚动和不安，但这些社会变迁也不能完全归罪于基础设施。即便对此人们努力反抗，但也无法阻止时代前进的步伐。时代要继续向前发展必须依靠更多的技术手段，通常要消耗更多的碳燃料，导致熵值不断增加。19世纪奥斯曼在巴黎组织了城市大改造。改造进行到了全盛阶段，当时的巴黎已经是满身疮痍，它曾是雨果《巴黎圣母院》这部作品的故乡。流离失所的维克多·雨果由此携众知识分子表达了他们看到历史悠久的巴黎遭到严重破坏

in 1920 one in five people had a car. This primacy was rewarded by the government's decision to establish a multi-billion-dollar freeway system of 66,000km begun in 1956. The expediency of its planners, urged on by the speculative interests of developers, laid waste to many central cities, such as Boston and Newark. In the effort to restrain the insensitive projects of so-called Urban Renewal a group of landscape designers, including Lawrence Halprin, prepared a primer, The *Freeway and the City* (1968), on how to work highways into the natural features of the landscape and to blend with the context of central cities. Halprin proved to be the most conscientious of this breed of road builders with his effort to mitigate the violence of the highways. At Freeway Park in Seattle (1972-1976) figure 2. Halprin spread a concrete tapestry of about 2ha over an 8-lane freeway that skirted the center of the city. He planted the park intensively with hundreds of tall trees, some of them sequoias, in deep concrete planters and paired them with roaring fountains, which work to neutralize the CO_2 and sound pollution of the traffic. Despite the violence and discomfort, not to say the social displacement, caused by the production and use of modern infrastructures, the efforts to resist them rarely match those to move forward. Progress usually implies the application of more technological interventions, and with that an increase in entropy, usually the incremental use of carbon fuels. In Paris during the heyday of Haussmann's transformations, many intellectuals, led by the exiled Victor Hugo, expressed their grief at the destruction of historic Paris, starting with the scraping of the Île de la Cité, the setting of Hugo's *The Hunchback of Notre Dame*. Individuals have often attempted to hold back the forces of eminent domain from destroying historic or natural sites, but only during the 1960s did they begin

图3 米洛大桥，诺曼·福斯特及合伙人事务所设计，法国，2004年
fig.3 Millau Viaduct by Foster + Partners, France, 2004

后心中产生的深切哀思。历史上有很多人站出来阻止一些决策者征用土地，以此保护一些历史遗址或自然景观免遭破坏。直到20世纪60年代，他们的努力才真正取得显著成效。在那样一个充满反对声音的年代，人们反对原子弹、反对种族歧视、反对越南战争，也因此掀起了反对高速公路和城市改造计划的热潮。美国方面，当时被誉为"建设大师"的纽约开发人员罗伯特·莫吉斯为了建设高速公路，试图破坏格林尼治村，简·雅各布斯等人纷纷站出来厉声抗争。这样的故事在多地上演，因此波士顿的北角区和旧金山的内河码头区才得以保留。雅各布斯的著作《美国大城市的死与生》在1961年问世，该书强烈倡议在城市发展的同时要维护社区居民尊严。威廉·怀特提出了对公共空间的一些理念，并因此获得一批追随者。他于1980年出版的《小型城市空间的社会生活》是倡导建设社交型都市理念的权威著作。欧洲方面，扬·盖尔在1971年出版（英文版于1987年出版）的《建筑之间的生活》中提到了哥本哈根的斯特罗里耶步行街。以这条长达1km的步行街为典范，他指出当代基础设施建设中的暴力性是可以被替代的。同一时期（20世纪70年代），荷兰也实行了"慢街道"（Woonerfs）规划项目，通过有目的地减小街道宽度来限制车辆速度，从而让出更多可供儿童嬉戏、居民休憩的空间。

反对粗暴型发展的先进理念虽然在各个地区，尤其是管理完善的中产居民区起到积极作用，但好景不长，巨大的现代主义浪潮涌进了城市，瓦解了密集的城市结构，破坏了自然环境，继续支配着整个城市空间的规划活动。当然，这也耗费了庞大的公共资金。从连接英国和法国的海底隧道的情况看，虽然投入了数十亿欧元的预算，但得到实惠的人只是极少数。在工程建设过程中，两

to achieve significant victories. The age of social movements against the atomic bomb, racism, and the war in Vietnam, inspired mass mobilizations against freeways and urban renewal projects. Protests such as the one led by Jane Jacobs to save her neighborhood in Greenwich Village from being destroyed by one of Robert Moses's freeways were repeated elsewhere, saving Boston's North End and San Francisco's Embarcadero. Jacobs's book *The Death and Life of the Great American City* (1961) still delivers a powerful message of how to maintain the dignity of neighborhoods against development. Defenders of public space rallied around the ideas of William H. Whyte, whose *The Social Life of Small Urban Spaces* (1980) became the Bible of socially oriented urbanism, while in Europe Jan Gehl in *Life Between Buildings* (1971, in English, 1987), expressed an alternative to the violent tendencies of modern infrastructure, using the Strǿget Car-Free Zone, a kilometer-long pedestrian street, in Copenhagen as a model. The movement for *Woonerfs* (slow streets) in Holland began during the same period in the 1970s, purposely narrowing urban streets to slow traffic down and give neighborhoods more spaces for children to play and people to sit.

While these high-minded forms of resistance had positive impacts on local districts, especially where there were well-organized middle-class neighborhoods, big modernism continued to conquer urban space in its efforts to break down the dense fabric of cities and invade natural environments, at great expense of the public coffers. The many billions spent for the Channel Tunnel, benefit only a small fraction of the populations of the UK and France, while requiring major intrusions on the landscape on both sides. The magnificent new bridge in the

图4 摩西项目，意大利威尼斯
fig.4 MOSE Project, Venice, Italy

个国家的自然风景都受到了严重破坏。图3是位于法国南部的米洛大桥，诺曼·福斯特爵士等人当初设计时，如果向西偏移30km，就能节省三分之二的预算，但他们还是为了宏伟的视觉效果，放弃了这个方案。这座气派的高架桥最终凭着它高耸的箭形桥塔，超过埃菲尔铁塔成为法国最高建筑。还有从2003年开始一直在施工的威尼斯摩西项目（图4），该项目建成后将在威尼斯泻湖口建造三座活动式水闸，在涨潮期拦住涌入城内的洪水。为此，政府投入了近60亿欧元，但其中15％被公务人员中饱私囊，其中35人因此被捕。这个计划于明年年初完工的项目不仅会扰乱附近海岸的生态系统，而且预计到2050年，亚得里亚海的海平面将会比防波闸门高出50cm，也就是说那时这个工程将寿终正寝。究竟是什么心态驱使着现代社会如此渴望这种粗暴式的进步，为什么不努力保持生活区域和自然环境的平静与安宁？这种心态可以类比人类首次发明并使用原子弹的心理，一旦得知技术方案有"可执行性"，就野心勃勃地将其等同为"须执行性"。或许，人生来就有强烈的破坏意识，它产生于人们的贪念和征服欲，人始终要与这种意识保持距离并纠正这种意识。有人辩解说迫于世界城市人口的不断增长的现实压力，人们无法不诉诸粗暴的解决手段。可这只是一种托词，我们可以认为这种压力其实是在呼吁人们从仿生学角度采取新型规划模式。想象一下在大城市孟买生活的情景，那里没有高速公路，平均通勤时间超过2.5小时；而在墨西哥城则有2500万居民面临水资源短缺和下水道堵塞的威胁。为了摆脱这样的困境，政府在所难免会下达命令采取激进的解决手段，可是这些手段多数都在依赖20世纪那种挥霍资源型的模式。其实，有一些合理技术手段可与当地管理相适应，采取这些手段会有助于社会背景下基础设施概念的提出以及环境友好型概念的形成。

south of France, the Millau Viaduct figure 3, designed by Sir Norman Foster and others, could have been placed 30km to the west saving about 2/3rds of the budget but would have resulted in a less spectacular project – two of its arrow-like pylons are now the tallest structures in France, surpassing the Eiffel Tower. The MOSE project in Venice figure 4, under construction since 2003, will place 3 mobile dikes to hold back the rising tides in the lagoon. Of the nearly 6 billion euro cost, 15% went to corrupt public officials, 35 of whom were eventually arrested. The system, scheduled for completion early next year, induces ecological damages to the biotopes of the lagoon, and will probably be obsolete by 2050, since the Adriatic is expected to rise half a meter higher that the level of the dikes. What is it that drives modern societies to desire such violent improvements rather than seeking to conserve the peace of their streets and natural environments? Much like the mentality that created and used the first atomic bombs, the potential that it "can be done" leads to the ambition that it "must be done". Humans seem to carry in their genes a highly destructive attitude of greed and conquest, one that constantly needs to be isolated and corrected. One could argue, however, that the pressure of the world's growing urban populations requires drastic solutions, but why must this be an excuse for further brutal interventions when it could be a new call toward biomimetic planning. Imagine living in a megacity like Mumbai, which does not have the advantage of fast roads and has an average commuting time of over 2.5 hours, or Mexico City where 25 million inhabitants live with the threat of water shortages and sewer backups. The mandate to do something fast is ineluctable, but too often the solutions rely on resource-squandering 20th-century models. The move toward

图5 巴塞罗那环形公路工程，特里尼塔·韦拉公园
fig.5 Barcelona's ring roads, Parc Trinitat Vella

在摆脱"路毙"式公共设施建设模式的例子中，最著名的则属20世纪90年代巴塞罗那城市的改造工程了。基础设施给这里的人们提供了各种附加的社会福利。为举办1992年的奥运会，巴塞罗那政府在整个城市进行了公共空间和基础设施的改造，为车辆和行人都提供了方便。为了缓解交通压力，在郊外开通了一条外环线路，同时还增设了12个地铁站点以及几条总长度为200km的自行车道作为补充。设计师没有选择建造破坏城市结构的隧道，而是选择了"随挖随填"的策略，因此空出来许多新的市政用地来建造体育场、社交中心、图书馆和公园。鲁伊斯·桑切斯构建的波布里诺公园景观，以及巴特列·伊·罗格打造的特里尼塔·韦拉公园景观，都是尝试缓解机动车流压力的很好的实例。尤为重要的一点是，巴塞罗那环形公路工程（图5）挪出资金来支持对10km长的海岸线进行清理的工作，成就了如今世界顶级的都市海滨景区，真实地响应了巴黎1968年5月风暴运动中一句具有乌托邦色彩的标语："铺路石下是海滩！"

意大利那不勒斯市政府为了修缮并扩展城市的地铁系统，邀请了著名的策展人阿吉莱·伯尼托·奥利瓦来负责指挥装饰工作。他找到几十位超前卫主义艺术家，让他们将绘画和雕塑融入系统的功能环境中，每一个车站都被赋予独特的风格。最后，由多米尼克·佩罗负责设计连接中央列车站的顶棚部分，也是最重要的部分，最终效果十分壮观。虽然在此之前倾注十年心血开展的挖掘工作必然会给人们留下心灵创伤，但这些伤口也在艺术的感染下慢慢愈合，地铁站也因此得到大家深深的喜爱。

我们能否借助改进机动车可移动性的名义来避免暴力破坏？欧洲几十个城市已经安装或恢复了电车系统，同时增加了自行车

appropriate technologies, linked to local management would favor a socially generated notion of infrastructure, and an environmentally friendly one.

Among the most famous alternatives to road kill public works were the transformation of Barcelona during the 1990s, where infrastructural projects became an easy access to collateral social benefits. The works planned for the 1992 Olympics led to a city-wide renewal of public spaces and infrastructures, providing convenience for vehicles and pedestrians. To ease the automobile traffic a ring road was completed around the outer neighborhoods, while a dozen new subway stops and 200km of bike paths were added as compensations. Instead of building tunnels, which disrupted the fabric of the city, the designers chose the "cut and cover" strategy, which yielded numerous new municipally owned sites transformed into sports fields, social centers, libraries, and parks. The landscapes by Ruiz Sanchez at Parc de Poblenou and Batlle i Roig at Parc Trinitat Vella are fine examples of attempting to soften the impact of automobiles. Best of all, the new project for the Barcelona's ring road [figure 5], or ronda, provided the finances to sanitize 10km of the coastline, which has become the world's best urban beach, literally realizing the utopian slogan launched in Paris during the revolts of 1968 – "Under the pavement is the beach".

Naples, in the effort to restore and expand its subway system asked the famous curator, Achille Bonito Oliva, to organize the decoration. His choice of dozens of trans-avant-guard artists, who integrated painting and sculpture into the functional surroundings of the system, gave a different identity to each station, concluding with

图6 水镜广场，法国波尔多
fig.6 Water Mirror, Bordeaux, France

图7 快速公交专用车道，巴西库里蒂巴
fig.7 Bus Rapid Transit(BRT) in Curitiba, Brazil

道。其中最成功的例子是法国波尔多的地面供电电车系统，到2009年已建成三条线路。沿线的许多车站（尤其是市郊地区的线路）都成为城市的观光景点。还有两个车站为市民提供了主要的休闲去处，其中排名第一的就是车站附近的水镜广场（图6）。它由米歇尔·高哈汝、彼埃尔·加涅和水利工程师让·马克斯·罗卡联合设计，于2006年建成，已成为欧洲最具人气的公共休憩场所之一。水镜广场与新建的有轨电车轨道平行，浅水池底部是光滑的花岗岩平面，占地面积超过3000m²，水面平静时，可以反射出18世纪的建筑——波尔多交易所的经典立面。白天，水雾开始从整齐摆列的管口中喷薄而出，仿佛云雾在地面缭绕。接着，管口渐渐变大，上面缝隙里涌出汩汩的水柱，哗哗作响地划破了水雾。随后，似潮汐交替一般，池中的水也神秘地逐渐消退。如此景致多引得大批的人前来嬉水欢闹，波尔多水镜广场因此成为一种新型的市内海滩。而在河对岸的巴斯提德车站附近，法国建筑师卡特琳·莫斯巴赫把废弃的调车场改造成了农作物观赏园，反映每个季节农作物的变化。通过给每个作物配备灌溉用的小型金属水池，保证每个季节生长不同的农作物，在植物园里形成随季节交替出现的迷人景观。

在建筑师兼市长杰米·勒纳的指导下，巴西的库里提巴市展示了一种更加温和的公共交通方式（图7）。这座城市位于巴西南部，拥有200万人口。政府没有为其挖掘地铁，也没有为昂贵的电车系统投入大笔资金，而是引进了5条快速公交专用车道。通过高效率中转站，这些车道履行着和地铁一样的职能。规划人员将车站作为带动社会投资的场所，吸引投资者来修建运动场、图书馆和公园。勒讷称之为"城市针灸术"。受到这种模式的启发，塞尔吉奥·法贾多让哥伦比亚的麦德林市从拉丁美洲暴力发展之首蜕变为和

the spectacular canopies designed by Dominique Perrault for the most important connection to the central train station. While the trauma of digging the subways for ten-year periods could not be avoided, the art has been a significant reward, much loved by the users.

Can we avoid the violent disruptions in the name of improving the alternatives to automobile mobility? Dozens of cities around Europe have been installing or restoring their tram systems, while adding bike lanes. One of the most successful was built in Bordeaux, three tram lines finished in 2009. Many of the stations, especially in the outer districts function as new urban attractors, while two of the stations have generated major spaces of leisure for the citizens: the first, the Water Mirror (*Miroir des Quais,* Michel Corajoud, Pierre Gangnet, and hydraulic engineer Jean-Max Llorca, 2006)[figure 6], has become one of the most loved public spaces in Europe. It stretches more than 3,000m² parallel to the new tram line on a smooth granite plane that when drenched in calm water reflects the 18th-century classical facades of Place de la Bourse. During the day, mist pours out of regularly placed holes forming a low-lying cloud, then water gurgles up through the cracks in ever greater jets to break the mist, after which the water mysteriously drains away much like the rhythm of the tides, attracting huge crowds who frolic in the water. Bordeaux's Water Mirror has become a new type of urban beach. Across the river at the stop of Bastide, a new Botanical Garden has taken the place of the abandoned switching yards. Designed by Catherine Mosbach to reflect the edible crops currently in season, the metal framed beds go through a fascinating crop rotation during the course of the year.

The Brazilian city of Curitiba, under the guidance of architect/mayor Jaime Lerner, demonstrated an even softer

图8 缆车作为城市公共交通手段，哥伦比亚麦德林
fig.8 cable cars as urban public transport in Medellin, Colombia

图9 阿马格·巴克垃圾焚烧厂，BIG设计，丹麦
fig.9 Amager Bakke incinerator by BIG, Denmark

平建设之都。从2004年开始，该市开通了许多极富创新性的中转线路，即在公交车难以到达的坡段配备空中索道缆车来输送乘客，同时在许多公交站点建设了新的公共空间、操场和图书馆（图8）。到了这一阶段，口号似乎该换一下了：基础设施如需建，公民福祉须兑现，我们相约海滩见！

最近还有丹麦BIG建筑事务所，也在实际建设中考虑了人们对海滩等类似公共空间的诉求。他们在哥本哈根建立了阿马格·巴克垃圾焚烧厂（图9）。一直以来，大多数城市垃圾都是运到填埋场填埋，有的甚至排放进大海，对海洋生物群落造成威胁。虽然最好的方式是通过回收再利用等方式减少垃圾，但这种垃圾能源厂提供了一种新型垃圾处理方式。他们在焚烧厂中设置催化转换器，把垃圾燃烧过程中的能源活用，整个过程几乎零污染。虽然没有人愿意在自家后院建设垃圾焚烧场，但阿马格·巴克垃圾焚烧厂却彻底改变了人们的这种观念。它在认真履行能源转换职能的同时，还能满足娱乐功能。这座焚烧厂看起来应该是哥本哈根市最高的建筑，设计师利用了这个优势，专门为攀岩爱好者设计了多孔状的建筑墙面。SLA建筑师事务所的设计师还在屋顶上种植了成排的树木，而且这个屋顶的平整度极为理想，所以冬天还会在上面做出一条滑雪坡道。

在规划过程中，有人采取了对环境更友好、更负责的态度，给基础设施赋予了改善环境的功能，利用基础设施减少二氧化碳等温室气体的排放。在过去的30年里，德国弗莱堡市在绿党组织的影响下，付出心血铺设了总长达400km的自行车道，还增设电车站点，保证所有居民从所在社区步行不到10分钟就能到达车站。同时，政府还鼓励太阳能产业的发展，支持寻找可替代能源的研究。

method of public transport figure 7. Instead of digging a metro for this city of 2 million in the south of Brazil, or spending a lot on expensive tram systems, five privileged express bus lanes were introduced, functioning like a subway, served by efficient transit stations. The planners used the stations as places of social investment, called by Lerner "Urban Acupuncture", for building playgrounds, libraries, parks. This model helped Sergio Fajardo transform Medellin, Colombia, from the most violent city in Latin America to become a reservoir of peace. Starting in 2004, the city introduced innovative transit lines to difficult-to-reach slopes with chairlifts, while preparing new public spaces, playgrounds, and libraries at many of the transit stops figure 8. The slogan should become: If infrastructure is inevitable, give us back some good public amenities: we want a beach!

The quest for a beach, or something analogous was recently put into practice for the new Amager Bakke incinerator in Copenhagen by the Danish architecture firm BIG figure 9. Most of the urban garbage of the world has been going into landfills, or worse dumped into the oceans where it endangers the maritime biomes. While the best practice would be to reduce the amount of waste through reuse and recycling, waste-to-energy incinerators, fit with catalytic converters, offer a new, nearly pollution-free solution to waste disposal. Historically no one desires an incinerator in their backyard, but the Amerger Bakke project is changing people's minds since it is as much an object of play as it is a serious transformer of garbage. The perforated elevations of what appears as the tallest building in the city are prepared for mountain climbing, while the roof has been planted by SLA Architects with banks of trees and during the winter will offer a ski slope for this famously flat region.

A more pacific attitude in the planning of infrastructure could lead to more ecologically responsible projects of

图10 弗莱堡总站，德国
fig.10 Freiburg main station, Germany

图11 苏德兰德公园，德国柏林
fig.11 The Südgelände park, Berlin, Germany

比如，总站的塔式结构部分是当时首批采用太阳能发电板作为表皮的建筑结构，而附近的一些居民停车场也设置了一些太阳能发电站生产清洁能源。另外，在车站附近建造的自行车停放处为了充分利用存储空间，采用了独特的圆形建筑以及相同的太阳能板屋顶（图10）。

20年前，被瑞士巴塞尔成为首个提倡建设绿色屋顶的城市，对所有大面积建筑屋顶的绿化工程都会给予部分资金支持。这种屋顶开始在城市被大范围地采用，能够在下雨时囤积50％的降水，吸收一部分二氧化碳和粉尘颗粒，仿佛就是一块大型的城市海绵。因而面对未来将会愈发严重的洪涝灾害，这的确不失为一种自然可行的办法。

很多城市也像波尔多市把调车场变成植物园这样，对废弃基础设施进行改造，形成一种反破坏式的发展趋势。另一个令人瞩目的例子就是位于柏林的苏德兰德公园（图11）。柏林市政府把坦普尔霍夫机场以西的废弃调车场改造成了公园，乍一看他们似乎在放任整个地带回归成莽原野林，但其实这些景观是经过25年的精心布设形成的，为的就是让森林覆盖满是废旧铁轨和机器等人工痕迹的地带，形成一种自然随性的风格。并且，公园里还设置了几道彩色墙壁，这个灵感来自墨西哥建筑师路易斯·巴拉甘，他把单色的墙面安插在机器与树木之间，产生了强烈的美学效果。

1997巴黎市政府规划出著名的艺术桥商业街（图12），它的原身是连接巴士底区域和里昂火车站的高架铁路桥。桥下半圆形拱门空间内隐匿着许多画廊和工坊，桥面则铺设了附有长凳的栈道，并种上了植物加以美化。更重要的是，巴黎的这座艺术之桥成为

conversion, turning infrastructure into a method for reducing CO_2 and greenhouse gases. Freiburg, Germany, under the influence of the Green party, over the past 30 years has conscientiously provided 400km of bike lanes, placing tram stops within a 10-minute walk of all neighborhoods, while encouraging the growth of solar industries and research on alternative energy. The tower at the train station was one of the first structures to use photo voltaic cladding, while the garages serving some of the neighborhoods are used to hold up solar farms, producing clean energy. A special round structure was built at the station for the efficient storage of bicycles and likewise has a solar roof.[figure 10]

Basel, Switzerland, initiated 20 years ago the first program for green roofs, partially subsidizing the planting of any large flat roof. The conversion of vast sections of the city with planted roofs, absorbs half of the rainwater that falls on it while also helping to neutralized CO_2 and fine particle pollution. To think of the structures of the city as part of a vast urban sponge offers a natural solution to a problem of flooding that will be growing intensely in the near future.

The conversion of abandoned infrastructures, such as the switching yards of Bordeaux into the Botanical Garden is beginning a trend to reverse the damage. The Südgelände park in Berlin[figure 11] offers a remarkable reuse of an urban void, and at first sight seems that if you let it grow the wilderness will return. But the landscape has been carefully orchestrated within the abandoned train yards west of Tempelhof Airport over the course of 25 years to feature a spontaneous forest that grew up amid the switching tracks and abandoned equipment. The area was installed with colored walls, inspired by the Mexican architect Luis Barragàn, placing monochromatic planes amid the machines and trees to great aesthetic effect.

图12 艺术桥商业街，法国巴黎
fig.12 Viaduc des Arts, Paris, France

图13 首尔路7017，MVRDV设计，韩国首尔
fig.13 Seoullo 7017 by MVRDV, Seoul, Korea

曼哈顿高架桥改建项目的灵感源泉。该项目由詹姆斯·科纳和迪勒·斯科菲迪奥&伦弗罗事务所共同负责，并由荷兰景观设计大师皮埃·奥多尔夫担任顾问。他们收集了大量野草来装饰老旧的高架铁路，使二者相映成趣。当地市民组织为此花了20年的时间进行准备，终于在2006年开始启动该项目并于2014年竣工。这条铁路现在由市政府和名为"高架线之友"的市民组织共同管理，正是由于该组织筹资建设，这座桥才能成为现在该城市的主要旅游景点之一。无独有偶，首尔也有类似的工程在2017年完成。荷兰设计团队MVDVR把一座废弃的高速公路高架桥改造成了一个空中行人花园，命名为首尔路7017（图13）。人行道长1km，645棵盆景树和24 000株花草星罗棋布，按韩国字母表顺序排列。

"建设速度要不断加快"，"最终效果要更加气派"，这些追求反映了20世纪基础设施建设者们飞扬跋扈的态度。我们很难说，持有暴力发展观念的人在将来是否会变得彻底"温驯"，但是经历这些之后，面对气候变化可能带来的种种不测，我们意识到我们应该发展更温和、更绿色、更平缓的基建模式，既能提升生活质量，同时又不伤害生态环境。否则，等待我们的将会是和巴拉德小说《撞车》中人物一样的下场，他们成为迷恋基础设施破坏力的病态偷窥者，眼睛盯着路面，等待下一场大事故的发生。尽管勒·柯布西耶之前大力宣传暴力性的基础设施，但年轻时的一次伊斯坦布尔之旅还是让他顿悟了。他领悟了一句格训："每建一条街，就栽一棵树。"对我们而言则应该更进一步：每建一条街，就造一片林！

In Paris the elevated rails connecting the Bastille area to Gare de Lyon were converted in 1997 into the Viaduc des Arts[figure 12], with galleries tucked into the ground level arches and wooden planks, benches, and plants above. This inspired the even more vital Highline project in Manhattan the team of James Corner, Diller Scofidio & Renfro, with consulting of the great Dutch landscaper Piet Oudolf, who regaled the old elevated rail line with a marvelous collection of wild grasses. The project was completed from 2006 to 2014 after nearly 20 years of organizing by citizen groups. It is now managed in conjunction with the municipality by Friends of the High Line, a citizen group that raised the money to fund what has become a major tourist attraction. In Seoul a similar conversion was completed in 2017, when the Dutch architects MVDVR transformed an abandoned elevated freeway in the Seoullo 7017[figure 13], a kilometer long pedestrian path dotted with 645 tree pots and 24,000 plants, arranged in a sequence inspired by the Korean alphabet.

It is doubtful that the aggressive attitudes exercised during the 20th century for ever faster, more dominant infra-structures, can ever be completely tamed. But in the aftermath, while we await the consequences of Climate Change we may start opting for softer, greener, slower infrastructures that improve social life while correcting ecological offenses. Otherwise we are doomed to the fate of the characters in J.G. Ballard's novel *Crash*, who become fetishistic voyeurs of the destructive capacity of infrastructures, waiting with their eyes fixed to the road for the next big accident. Le Corbusier, despite all of his propaganda for destructive infrastructure, had an epiphany while visiting Istanbul as a young man, learning the dictum: "When you build a street plant a tree." We should take this further and demand a forest!

芬兰首都赫尔辛基市区地铁西线延长线一期建设已经竣工，于2017年11月投入运营。

西线连接赫尔辛基市的草湾区至埃斯波市的马丁镇，每天客流量超过17万人。该项目一期建设了八个站点，二期计划增设五个站点（预计在21世纪20年代早期投入使用），其设计宗旨是使车站在城市和室内设计两个方面都能彰显其独特的建筑风格和地方特色。

与此同时，这条线路将连接埃斯波市的多个中心区域，并以此建立全新的城市主干，显著地提高城市交通网的活力。ALA建筑师事务所和Esa Piironen建筑师事务所合作建设了一期项目中的两个新站点：阿尔托大学站和Keilaniemi站。

自20世纪五六十年代起，位于该市的奥塔涅米校区就怀着对未来的憧憬，运用科技手段，使自身始终保持着鲜明的地方特色，也因此成为最能展现芬兰建筑魅力的地方之一。2010年，原来的赫尔辛基理工大学、赫尔辛基艺术设计大学和赫尔辛基经济学院三所学校合并，最终成立了阿尔托大学。设计师阿尔瓦·阿尔托于20世纪50年代开始负责校园的总体规划，设计出成为校园标志性景观的红砖建筑。阿尔托大学地铁站就位于校区的核心地带，其主入口正对前赫尔辛基理工大学的主楼。

在众多的设计决定中，阿尔托大学站凭借在设计上运用了丰富的材料，与沿线的其他地铁站明显区别开来。设计师没有选用色彩绚丽的涂料，而是着力表现材料的天然质感。天花板也被放低，材料采用耐候钢板，其颜色不仅在视觉上与地铁站所有的公共区域保持一致，还同周边建筑的砖红色相互映衬。由多个小平面拼接而成的天花板从主入口一路延伸至站台层，并在上方与位于Tietotie大街的次入口相连。

入口处地上部分的建筑材料主要由三部分构成——略显陈旧的铜覆层、灰色花岗岩和耐候钢板。入口空间的结构简约轻盈，动感十足，让人联想起折纸艺术。

自然光通过自动扶梯井照射到站台层，在站台层可以看到奥塔涅米大厦的石灰走廊。

阿尔托大学地铁站
Aalto University Metro Station

ALA Architects + Esa Piironen Architects

The first phase of the western extension of the Helsinki metropolitan area subway line, the West Metro, was completed and opened for traffic in November 2017.
Connecting Ruoholahti, Helsinki to Matinkylä, Espoo, it services over 170,000 passengers per day. The objective that has been set for the architecture of the eight new stations along the first phase and the five new stations along the second phase (to be opened for traffic in the early 2020s) of the extension was to create distinctive, location-specific identities in both urban and interior scale.
At the same time, the metro aims to act as a link between the various urban centers of the City of Espoo and in a way creates its new backbone, remarkably changing the dynamics of the network-like urban structure of the city. Among the new stations in the first extension phase, the mission given to ALA Architects and Esa Piironen Architects was to

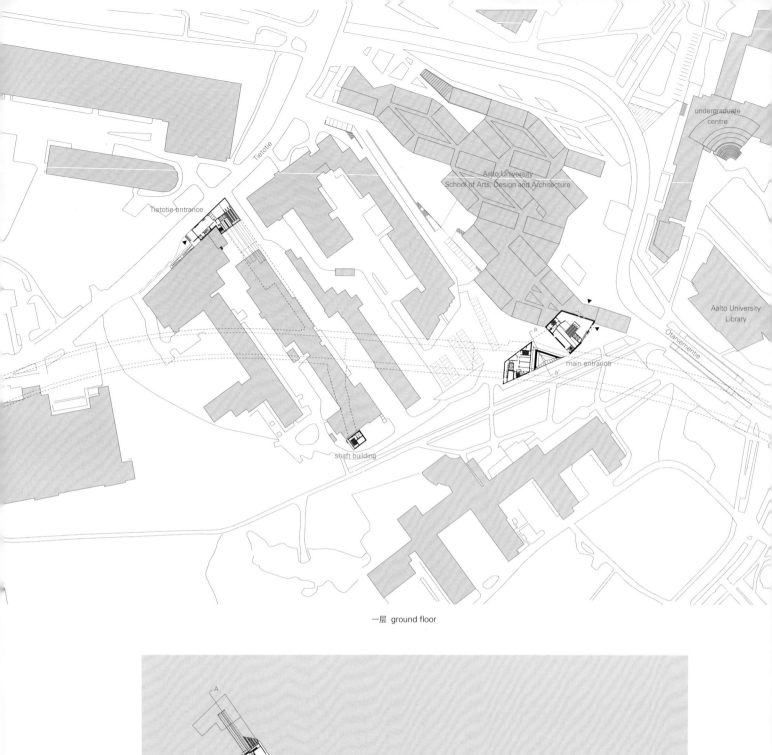

一层 ground floor

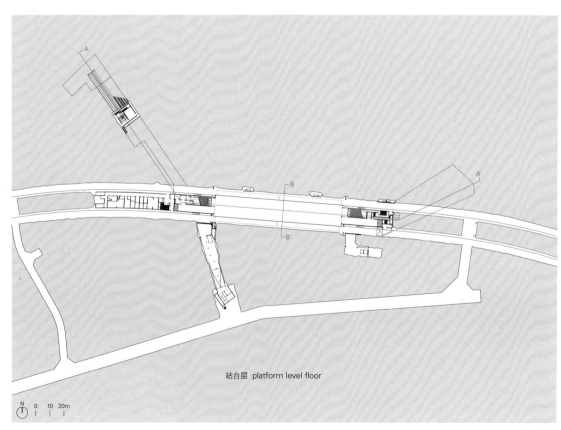

站台层 platform level floor

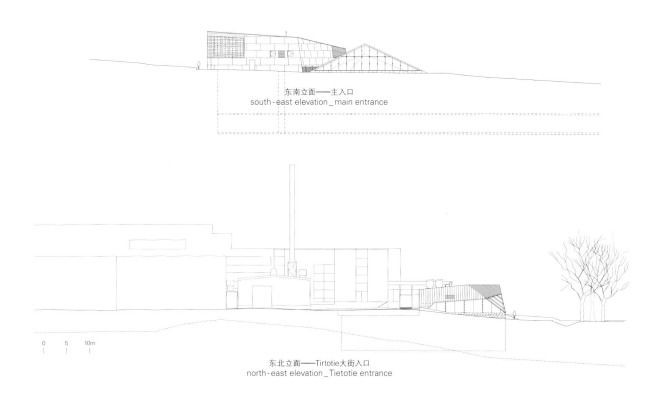

东南立面——主入口
south-east elevation_main entrance

东北立面——Tirtotie大街入口
north-east elevation_Tietotie entrance

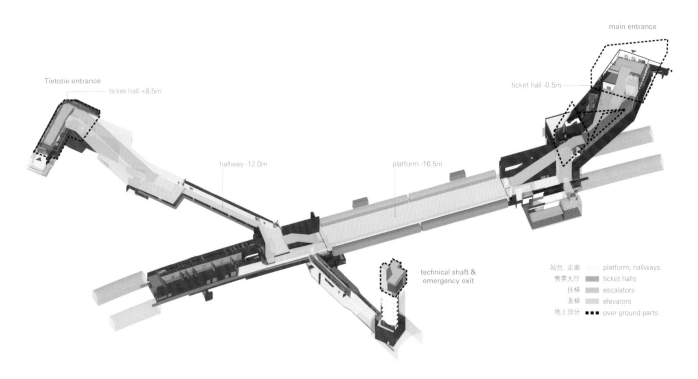

项目名称：Aalto University Metro Station / 地点：Espoo, Finland / 事务所：ALA Architects + Esa Piironen Architects / 项目团队：ALA Architects–Juho Grönholm, Antti Nousjoki, Janne Teräsvirta (until 2015), Samuli Woolston, Anniina Koskela, Harri Ahokas, Santtu Hyvärinen, Niklas Mahlberg, Olli Parviainen, Jorge Rovira, Pekka Sivula, Pekka Tainio, Jyri Tartia, Yena Young; Esa Piironen Architects–Esa Piironen, Juha Lumme, Henriikka Ryhänen / 项目顾问：Sweco PM / 协调者：CJN Arkkitehtitoimisto / 地理、铁道、岩石工程师：Konsulttiyhteenliittymä FKW / 结构设计：Insinööritoimisto A-Insinöörit
暖通空调工程师：Pöyry Building Services, Insinööritoimisto Olof Granlund and Konsulttiryhmä Nissinen-Niemistö / 车站总承包商：YIT Rakennus
供应商：Metek–glass surfaces; Forssan Sisärakenne–lowered ceiling; Kone–elevators and escalators; Seroc–fiber concrete / 客户：Länsimetro
用途：underground metro station with two entrance pavilions and a technical shaft building / 总楼面面积：15,500m² / 委托时间：2009 / 竣工时间：2017 摄影师：©Tuomas Uusheimo (courtesy of the architect)

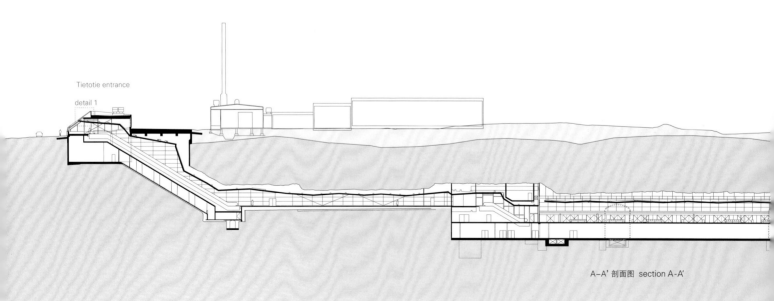

A-A' 剖面图 section A-A'

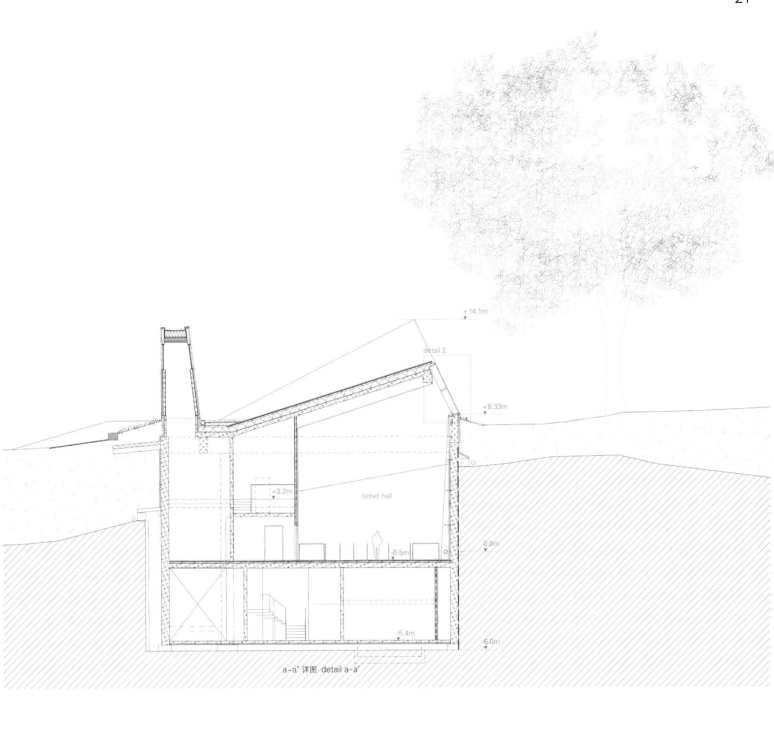

a-a' 详图 detail a-a'

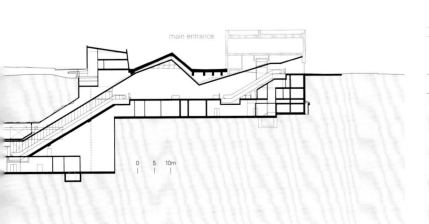

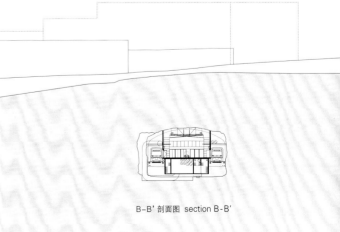

B-B' 剖面图 section B-B'

collaboratively design the two – the Aalto University Station and the Keilaniemi Station.

Since the 1950s and 60s, Otaniemi is a district in Espoo with a strong individual character linked to science, technology, and the general belief in the future, thus becoming one of the most interesting sites of Finnish architecture. In 2010, it became a home to Aalto University, which is formed by merging the Helsinki University of Technology, the University of Art and Design Helsinki, and the Helsinki School of Economics. The campus masterplan and its representative red brick buildings were designed by Alvar Aalto in the 1950s. At the heart of this campus, the Aalto University metro station opens its main entrance straight towards the former Helsinki University of Technology main building. Among other design decisions, the station distinguishes itself from the other stations along the metro route through its rich material palette. The palette has been picked to avoid a high gloss, but rather to emphasize a natural materiality. The station's lowered ceiling is made of Corten steel panels. Its color not only visually connects all the public areas of the station, but also relates back to the surrounding red brick environments. The faceted ceiling flows through the main entrance, to the platform level and up to the secondary entrance on Tietotie street.

Aged dark copper sheet cladding, grey granite, and Corten sheets form the basis of the material palette for the above-ground parts of the entrance pavilions. Visible structures are reduced in the entrance space with an engineered, athletic, light form, reminiscent of an origami.

Natural light is brought down to the platform level via an escalator shaft that terminates with a view towards the lime alley of the Otaniemi Mansion.

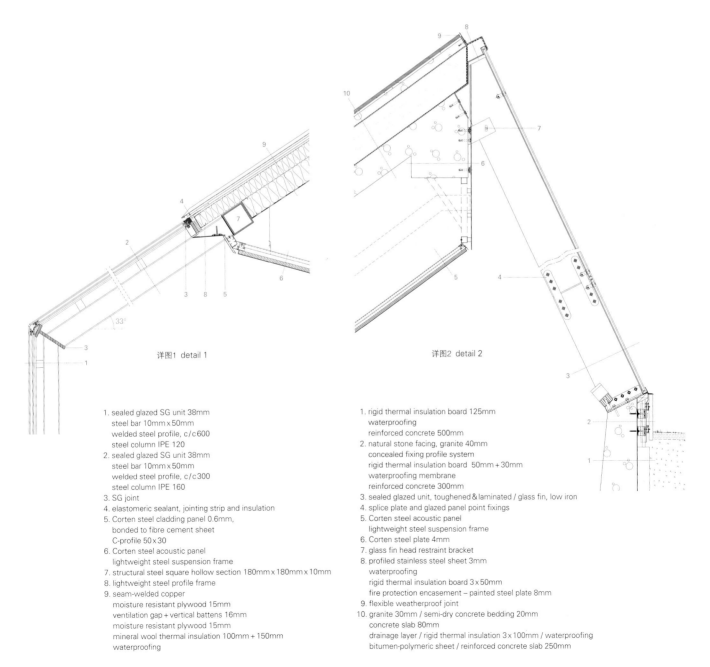

详图1 detail 1

详图2 detail 2

1. sealed glazed SG unit 38mm
   steel bar 10mm x 50mm
   welded steel profile, c/c 600
   steel column IPE 120
2. sealed glazed SG unit 38mm
   steel bar 10mm x 50mm
   welded steel profile, c/c 300
   steel column IPE 160
3. SG joint
4. elastomeric sealant, jointing strip and insulation
5. Corten steel cladding panel 0.6mm,
   bonded to fibre cement sheet
   C-profile 50 x 30
6. Corten steel acoustic panel
   lightweight steel suspension frame
7. structural steel square hollow section 180mm x 180mm x 10mm
8. lightweight steel profile frame
9. seam-welded copper
   moisture resistant plywood 15mm
   ventilation gap + vertical battens 16mm
   moisture resistant plywood 15mm
   mineral wool thermal insulation 100mm + 150mm
   waterproofing

1. rigid thermal insulation board 125mm
   waterproofing
   reinforced concrete 500mm
2. natural stone facing, granite 40mm
   concealed fixing profile system
   rigid thermal insulation board 50mm + 30mm
   waterproofing membrane
   reinforced concrete 300mm
3. sealed glazed unit, toughened & laminated / glass fin, low iron
4. splice plate and glazed panel point fixings
5. Corten steel acoustic panel
   lightweight steel suspension frame
6. Corten steel plate 4mm
7. glass fin head restraint bracket
8. profiled stainless steel sheet 3mm
   waterproofing
   rigid thermal insulation board 3 x 50mm
   fire protection encasement – painted steel plate 8mm
9. flexible weatherproof joint
10. granite 30mm / semi-dry concrete bedding 20mm
    concrete slab 80mm
    drainage layer / rigid thermal insulation 3 x 100mm / waterproofing
    bitumen-polymeric sheet / reinforced concrete slab 250mm

先锋村站
Pioneer Village Station
aLL Design

在多伦多市交通委员会（TTC）的指示下，加拿大多伦多—约克士巴丹拿线地铁扩建项目大力启动，计划加建六个全新的地铁站。该项目由建筑师威廉·艾尔索普带领的aLL Design工作室和士巴丹拿公司（TSGA）共同负责，目前已完成其中两个站点的设计和施工。士巴丹拿公司由IBI集团、LEA咨询有限公司及科进建设公司合资建立。

多伦多—约克士巴丹拿地铁延长线于2017年通车，它贯通了多伦多市区和约克区的交界地带，成为多伦多市现有地铁系统的重要延伸线路。

先锋村站位于约克区的边界上，地处斯蒂尔斯大道西与西北门户的交会点，占据了约克大学校园的一角。该站点将作为这个综合性区域的交通枢纽，每天为2万名地铁乘客服务，并提供1881个通勤车停车位和两个独立的公交总站。同时，地铁站入口和公交终点站将形成一个新的公共联络点，给以斯蒂尔大道西为起点的周边不发达地区的日后兴修预留了通道空间。

威廉·艾尔索普说他们从教堂建筑中找到了很多灵感。形成地铁站入口部分的两座建筑物雕塑感十足，且高得出奇，即使在很远处也依然可见。两者在结构上采用耐候钢材，形状大小也非常协调。车站的顶棚也是用耐候钢制作的，它采用了一个大型悬臂式屋顶设计，屋顶上面还铺着草坪。这个"绿色屋顶"可以为候车的乘客提供一个很好的遮阳场所。利用绿色屋顶景观和其他建筑径流区周围的绿化带，可以减少雨水流入市政排水系统。除了绿色清凉的屋顶之外，这地铁站多处践行了多种环保理念：增强采光以减少照明用电；以LED灯作为电缆塔标识的照明设备，指路标识的照明采用一些节能装置以降低建筑能耗。

先锋村站采用了精心设计的混凝土制作工艺，内墙采用经过高抛光处理的混凝土材料。沿站台横向排列的柱子也都与地面保持一定斜角，形成椭圆形的截面。

这种"超级雕塑"的概念也体现在了地铁站站台层。来自柏林的reality;united工作室提出了"光影魔法"的设计理念，将地铁站照明系统与光影艺术完美结合，形成一种互动式装置。这种装置由发光构件构成，每40个发光构件为一组，形成字母、特殊字符、数字符号等文字信息，悬挂在整个地铁站的天花板上，使每个乘客都能够看到显示在屏幕上的信息。

Will Alsop's practice, aLL Design, in collaboration with The Spadina Group Associates (TSGA), a joint venture of IBI Group, LEA Consulting Ltd. and WSP, have designed and completed two of the six stations on the Toronto Transit Commission's (TTC's) new TYSSE (Toronto-York Spadina Subway Extension).

The Toronto-York Spadina Subway Extension Project is open in 2017 to provide a critical extension for the existing TTC subway system across the municipal boundary between the City of Toronto and The Regional Municipality of York. Pioneer Village Station straddles the border of York Region, beneath the intersection of Steeles Avenue West and Northwest Gate, anchoring a corner of York University Campus. The station will serve as an integrated regional transport hub serving up to 20,000 subway passenger trips daily, providing 1,881 commuter parking spaces and two separate regional bus terminals. Meanwhile, the station entrances and bus terminals will create a public focal point that will

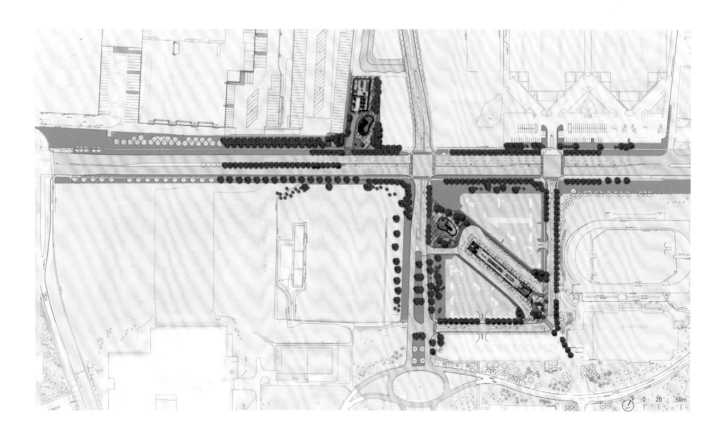

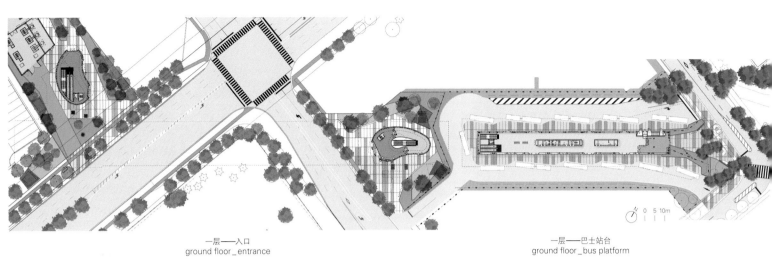

一层——入口
ground floor_entrance

一层——巴士站台
ground floor_bus platform

地下一层——北服务间
first floor below ground_north services room

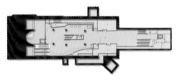

地下一层——南步道
first floor below ground_south walkway

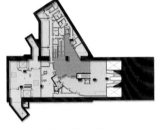

地下二层——北大厅
second floor below ground_north concourse

地下二层——南大厅
second floor below ground_south concourse

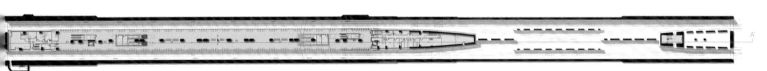

地下三层——站台
third floor below ground_platform

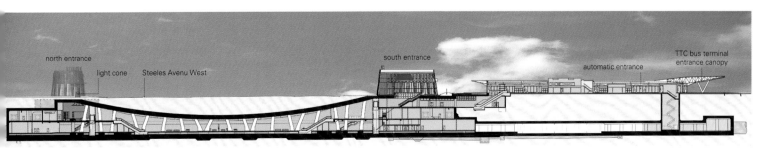

A-A' 剖面图 section A-A'

项目名称：Pioneer Village Station / 地点：Toronto, Ontario, Canada / 事务所：aLL Design / 项目团队：Will Alsop, Anaïs Bléhaut, Melanie Clarke, Dieter Janssen, Sonila Kadillari, Christina Kalt, Vincent Lin, Duncan Macaulay, Steve Mason, Tarek Merlin, Ed Norman, Maxine Pringle, Philip Richards, Arnold von Storp, George Wade, Greg Woods, Bonny Yu / 参与设计建筑师：IBI Group - Bruce Han, Richard Stevens, Celia Johnstone, Charlie Hoang, Ana-Francisca de la Mora, Gui Chan, Stuart Hill, Michael Norton, Welland Sin, Michael Mueller, Jennifer Ujimoto, Colleen Gono, Amer Obeid, Andrew Chiu, Ashley Adams, Adetokunbo Bodunrin, Bijan Ghazizadeh, Bill Whitelaw, Claudia Rosario, Domenico Grossi, Gary Chien, John Lenartowicz, Timothy Mitanidis, Trevor McHugh, Manisha Athavale, Kirbi Abuyan, Nasir Jaffer, Jim Bazios, Shailza Bhavsar, Behrang Ghamisi, Marjan Zelic, Keerthana Balakunalan, Gabriel Colombani, Syed Navqi / 艺术家：Bruce McLean, Realities:United (Lightspell), Jan Edler, Tim Edler / 土木、结构工程师：LEA Consulting Ltd. 结构工程师：WSP (Halsall Associates) / 机电管道工程师：HH Angus & Associates Ltd. / 景观设计：Janet Rosenberg & Studio Inc. 客户：Toronto Transit Commission / 用途：transportation – metro and bus stations / 背景：major interchange on newly extended Spadina Line 总楼面面积：16,200 m² / 造价：$165M CAD / 设计时间：2011 / 竣工时间：2017.12 / 摄影师：©Wade Zimmerman (courtesy of the architect)

serve as a catalyst for the future development of the currently underdeveloped surrounding area, beginning with Steeles Avenue West.

"We learn a lot from cathedrals." said Will Alsop. The subway station's entrances were designed as a pair of sculptural structures; their height exceeding the necessity, to increase the visibility. Rendered in weathering steel, these two structures mirror each other in shape and scale. The bus station canopy – also of Corten steel – has a huge cantilevered roof planted with meadow grasses, to create a "green roof" and to provide shelter for waiting passengers. Utilizing green roof landscaping and soft landscaping areas adjacent to other buildings' runoff areas allowed for a reduction in stormwater runoff into the municipal drainage system. Other than cool roofs and green roofs, the station incorporates many environmental initiatives such as increased daylight

levels to reduce electric lighting power usage, LED lighting in pylon signs, and energy-efficient lighting in illuminated wayfinding signage to reduce power consumption. Pioneer Village station comprises beautifully executed concrete work – the interior walls are highly-polished concrete and the supporting columns along the length of the platforms are angled and ovoid in section.

The idea of "super sculpture" also resides inside the station, on the underground platform level. "LightSpell", designed by realities:united, Berlin, is a hybrid between art and the lighting of the subway station. The interactive installation consists of a suspended array of 40 light elements that run along the ceiling throughout the station. Each element produces the alphabet, special characters and numerals, enabling passengers to receive messages that appear on the display.

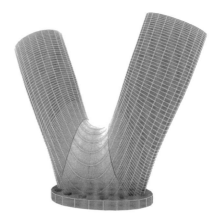

普林斯顿车站大厅和商店
Princeton Transit Hall and Market

Rick Joy Architects

普林斯顿车站大厅是Rick Joy建筑师事务所设计的第一个公共项目。铺着黑色钢板的屋面呈尖屋顶设计，界定出了一个细长的体量，这个体量中容纳了候车大厅，给项目设定了亦动亦静的基调。为了使车站与周围的环境融为一体，建筑师们有意把支撑屋顶的柱子设计成形状高挑的长柱。这种设计使车站大厅和校园建筑之间产生了视觉联系，不仅如此，无论是站内还是站外，宽敞明亮的空间布局都给人一种肃静宏伟的感觉，与普林斯顿大学校园内许多哥特式建筑相映成辉。

该项目设计方案简洁明了，就是在中心广场两侧规划两个不同大小的体量，人们可以在广场中间通行或逗留。黑色钢板在视觉上起到了将两个体量连接起来的作用。候车大厅的体量正好位于屋顶坡度陡降的那一部分，屋面内侧向地面倾斜的坡度很长，在来自南侧的光线照耀下，形成细长的、温暖的光带。单独建一个候车大厅是根据设计任务书而做出的深思熟虑的决定，设计任务书要求同时建造多个不同规模的空间，比如建一个卫生间和一家便利店。

通过将候车室与外界环境分隔开来，建筑师创造了一个隔绝了视觉噪声和听觉噪声的崭新空间，候车的旅客们可以在此享受片刻的清静。候车室内摆放着由木制品工艺设计师中岛乔治设计的黑胡桃木长椅，接近落地窗高度的窗户都朝向南面，被白橡木边框包裹，这些设计给整个室内平添一种温暖的感觉。室外方面，不同尺寸的沙灰色混凝土柱遮挡了来自北面的强光和列车驶过的噪声，同时其富有律动感的线条将建筑与广场对面的空间连接起来。这个空间呈L形，里面设有卫生间和Wawa便利店。在这里，光滑的灰色混凝土与黑色的钢板窗间壁结合在一起，搭建出一条有顶棚的人行道，并将人们的视线吸引至头顶，在那里，它们折叠形成绿色屋顶，宽阔而青翠的空间平衡了建筑庄重、

低矮的造型。看完屋顶设计我们回过头来观察一下地面的设计，从候车室内部向外延伸，进入中央广场，并穿过Wawa便利店，在景观设计师们的精心打造下，地面的铺筑模式呈现出富有活力却又一丝不苟的景观效果。在中心广场，纤细的树木提供了充足的遮阳空间供旅客在正午十分稍坐、小憩。

该项目吸收了校园蓬勃的朝气，创造出动静皆宜的空间。当阳光透过整齐的窗口洒到室内空地上时，沐浴在斑驳的光影下的人们，哪怕做着像等车或买杯咖啡外带这样再简单不过的事，竟然也会神奇般地感觉到心旷神怡。

At Princeton Transit Hall, the first public project by Rick Joy Architects, a sharp, peaking roof form rendered in blackened steel, defines a long, slender volume that houses a waiting hall and sets the tone for a project that is at once vibrant and serene. Strategically, the tall, slender columns that support the roof make the building feel at home in its surroundings. But it is more than just the visual effects of the columns that establish a link between the transit hall and the architecture of the campus; the ample, luminous feeling of the spaces, both inside and outside, has the quiet grandeur that characterizes many buildings on Princeton's Collegiate Gothic campus.

The project's plan is simple: two volumes housing two different levels of activity flank a central plaza where people

项目名称：Princeton Transit Hall and Market
地点：Princeton, New Jersey, U.S.A.
事务所：Rick Joy Architects / 主创建筑师：Rick Joy
高级合伙人：Matt Luck / 项目团队：Bach Tran, Natalia Hayes, Philipp Neher, Luat Duong, Shawn Protz, Heiman Luk, Claudia Kappl (Lighting)
客户：Princeton University Board of Trustees
操作者：Princeton University, New Jersey Transit, Wawa
施工经理：Turner Construction Company
规格制定者：Construction Specifications, Inc. / 标识设计：Two Twelve
结构工程师：Arup Engineering
照明设计：Arup Engineering with Rick Joy Architects
规范咨询：R.W. Sullivan Engineering / 绿色屋顶设计：Roofmeadow
土木工程师：Vanasse Hangen Brustlin / 成本估算：Davis Langdon
景观设计：Michael Van Valkenburgh and Associates
总承包商：Turner Construction Company
建筑类型：transportation, institutional, retail
功能：train station, canopy, marketplace, bike storage
用地面积：89,030.84m² / 建筑面积：1,412.12m²
结构：cast-in-place concrete, steel structure, concrete masonry
室外饰面：precast concrete forms, blackened stainless steel, cast-in-place concrete, bluestone floor paving
屋顶：blackened stainless steel roofing, extensive green roof, PVC roof membrane
洞口：white oak framed doors and windows, aluminum-framed doors and windows, interior wood doors, custom skylight, hollow metal doors and frames, glazing
室内饰面：American Black Walnut benches, blackened stainless steel ceilings, paints, stains, coatings
设计时间：2012—2013 / 施工时间：2014—2018
摄影师：©Jeff Goldberg/ESTO (courtesy of the architect)

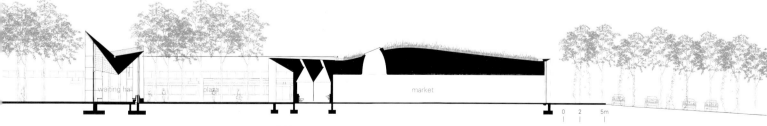

A-A' 剖面图 section A-A'

can pass through or linger. The use of blackened steel connects the two volumes visually. The waiting hall volume manifests the form of a roof volume that plunges deep inside, with its underbelly resting over a long space into which warm southern light washes in swaths. The decision to create a stand-alone waiting hall was an intentional solution to the project brief, which called for programs of varying intensity, including a bathroom and a convenience store.

By separating the waiting room, the architects created a pristine space, free from visual and aural noise, for travelers to pause before boarding their train. In the waiting room, black walnut benches designed by George Nakashima Woodworker and white oak framing the tall south-facing windows lend the space a sense of tactile warmth. On the exterior, the varyingly-sized sandy-colored concrete columns block harsh northern light and noise from the train tracks while establishing a dynamic rhythm that connects the structure to the L-shaped volume, which houses the bathrooms and the Wawa convenience store, across the plaza. Here, smooth gray concrete combines with blackened steel piers that delineate a covered walkway and draw the eyes up toward the point where they fold to create a green roof, the verdant expanse balancing the structure's demure, low-slung profile. Back on the ground, a lively yet rigorous paving pattern, produced in collaboration with the landscape architects, extends outward from the interior of the waiting room, into the central plaza, and across to the Wawa. In this central space, slender trees provide just enough shade to sit underneath them for a midday break. The project borrows the campus's dynamism to produce spaces capable of capturing moments of stillness just as easily as they make room for fast-paced activity. As light washes inside through rhythmically spaced openings, simple acts like waiting for the train or buying a cup of coffee to-go, become magical.

拉赫蒂枢纽站
Lahti Travel Center

JKMM Architects

新枢纽站位于芬兰南部城市拉赫蒂市的中心,旁边是历史悠久、至今仍在运营的火车站。新枢纽站作为一个公共交通枢纽,连接着铁路系统与长途线和本地线的公交系统。

新枢纽站包含一个60m长的棚式公交终点站台、闭合的电梯结构、本地公交停车点和配套景观。站台下还有一条80m长的封闭隧道。这些元素在不同层面上为复杂的城市环境创造了一个易感优质的实体空间。

旁边的火车站有着重大的历史意义,其历史可以追溯到20世纪30年代,这座红砖建筑为新枢纽站奠定了个性色彩浓厚的环境基调。火车站已被纳入芬兰国家文物局的国家重点文化环境名录中。矗立在它部分楼体前面的棚式建筑是新建的城际列车的终点站。终点站的顶棚作为新交通枢纽的最突出要素,完美融合了新旧元素,它那极尽简约的雕塑外观彰显了该区域的历史意义和重要价值。

就城市景观而言,新枢纽站更是品质优良,给人以整体性极强的视觉冲击。为了让建筑更好地与周边环境相融合,建筑师师精心挑选出铜、铝和玻璃作为建筑材料。顶棚和柱子采用穿孔铜板;旁边的塔式电梯在外墙与承重构件上都使用了玻璃,造型精致,通风良好;玻璃里面的电梯井被铜板和铜缆包住。在建筑师的打造下,电梯塔造型的柔美与车站顶棚流线型轮廓的魄力相得益彰。其他两部电梯采用了同样的玻璃和铜材质,设立于北部区域,连接了底层的街道与北侧Mannerheiminkatu街的公交站。这些电梯塔成为前往城市南部的重要通道。

桥下空间的覆层采用阳极氧化铝型材。降噪设备、普通照明设备和用来烘托气氛的高级照明设备都被整合到覆层的后面,给予了整个隧道空间优雅惬意的视听效果。主体结构之间的支撑墙体、桥栏杆、室外座椅和外墙等部件同样全部由铜板覆盖,以加强外观的整体性。

如今枢纽站已开始全年昼夜不停地运营,因此照明是一个重点。电力和暖通空调设备隐藏在建筑内部,光源也藏在不同构件的穿孔铜板覆层后面,在天色暗淡的时段里可以增强铜板的质感。

The new Travel Center is located at the heart of the city of Lahti next to the existing, historical railway station. It forms a public transport hub connecting the rail network to the long-distance and local bus lines.

The building consists of a 60-meter-long canopy for the bus terminal, enclosed elevator structures, local bus stops, and supporting landscape elements. There is also a 80m long tunnel space underneath the new deck. Together these elements create an easily perceivable and high-quality entity in a complex city environment at various levels.

Historically significant railway station, a solid red brick building from the 1930s, sets a characteristic milieu for the Travel Center. The station building is included in the National Board of Antiques' list of nationally significant cultural environments. Partially in front of it stands the new terminal canopy for the intercity buses. As the most prominent element of the new Travel Center, the terminal canopy initiates a dialogue between the new and old elements. Its minimal-

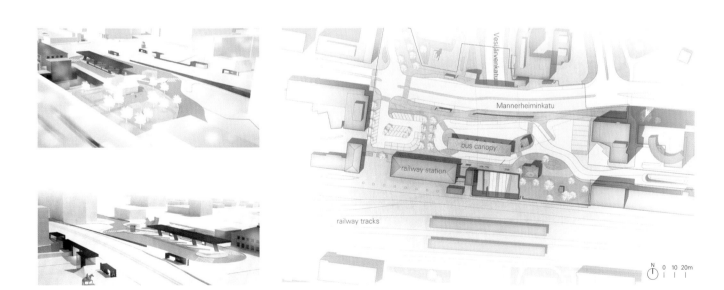

项目名称:Lahti Travel Center / 地点:Lahti, Finland / 事务所:JKMM Architects / 项目团队:Samuli Miettinen-chief designer; Tuomas Raikamo-project architect in charge; Katariina Knuuti-architect; Marko Pulli-architect, 3D-modelling, visualization; Jarno Vesa-interior architect, 3D-modelling, visualization; Asmo Jaaksi, Teemu Kurkela, Juha Mäki-Jyllilä-architect, partner / 城市规划、城镇景观:City of Lahti, Zoning Department / 文化环境:City Museum of Lahti / 结构设计与附属结构设计:Sito Oy / 总体规划与交通设计:Trafix Oy / 建筑设备工程师:Projectus Team (now Ramboll) / 施工:YIT Oy / 合作伙伴:VR Group-National Railway Company; Matkahuolto-National Bus Transport Cooperative; Linja-autoliitto-National Association of Bus Traffic Companies; Liikenneministeriö-Ministry of Transport and Communications; Näkövammaisten keskusliitto-Central Association of Visually Impaired; Vammaisneuvosto-Municipal Board for Disabled People / 主要分包商:UPPE-glass; Three L Technologies-copper cladding; Arston-aluminium cladding; Lahtinen & Kumppanit-steel fittings; Rinaldo-railings and handrails; Graniittikeskus-stone claddings / 客户:City of Lahti, technical and environmental branch / 用途:terminal canopy, 3 elevator and stairwell towers, cladding for the underdeck space, 3 bus stop shelters, cladding for the spatial structures, outdoor benches, display cabinets and other outdoor furniture / 建筑面积:11,000m² / 设计开始时间:2011.12 / 施工时间:2014.4—2016.2 / 摄影师:©Mika Huisman (courtesy of the architect)

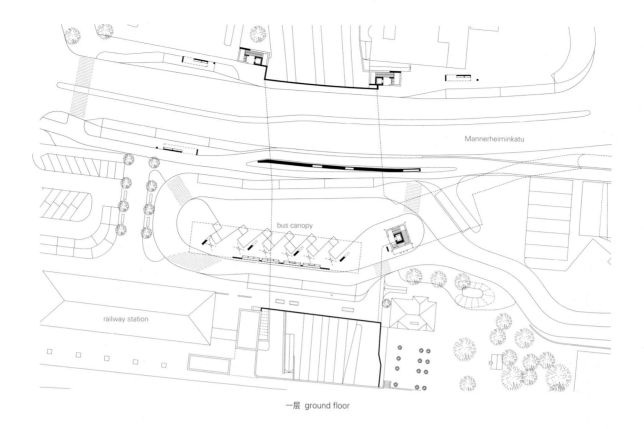

一层 ground floor

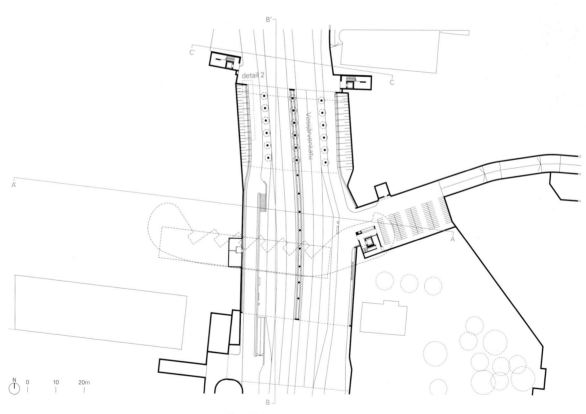

地下一层 first floor below ground

A-A' 剖面图 section A-A'

B-B' 剖面图 section B-B'

C-C' 剖面图 section C-C'

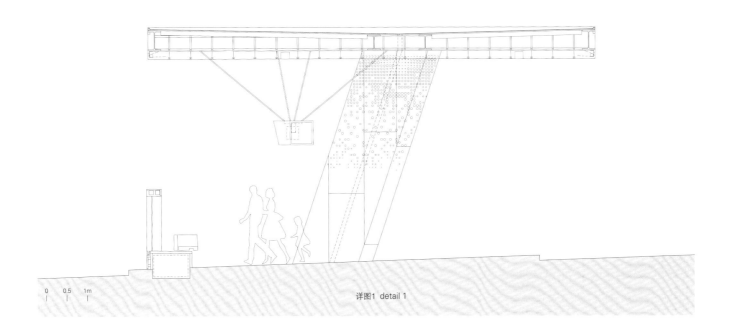

详图1 detail 1

istic sculpture-like form embraces the history and value of the area.

In terms of the cityscape, the Travel Center provides a high quality and cohesive visual impression. The main materials, copper, glass and aluminum, were carefully chosen to meet requirements of the surrounding milieu. The canopy and columns are clad in perforated copper. Next to it, the delicate and airy elevator tower uses glass in both the outer walls and load-bearing structures. Inside the glass shell is the elevator shaft, covered in copper and copper wire mesh. It is an elegant counterpart to the powerful and streamlined silhouette of the canopy. The two other elevator towers, also made of glass and copper, are located in the northern part of the area. The elevator towers connect the lower level street to the northern bus stop shelters on Mannerheiminkatu. Together they become a portal to

southwards part of the city.

The space under the bridge deck is clad with anodized aluminum profiles. Noise reduction, general lighting and high-quality atmospheric lighting are all integrated behind the cladding. They create a visually refined and acoustically pleasing environment to the tunnel-like space. The parts between the main structures – support walls, bridge railings, outdoor benches and walls – are all copper-clad as well to complement the cohesive appearance.

The Travel Center is in use throughout the year and around the clock. Therefore special attention was paid to lighting. The electricity and HVAC equipment is hidden inside the structures. Light sources have been placed behind perforated copper parts in various elements and will enhance the character of the copper parts during the darker seasons.

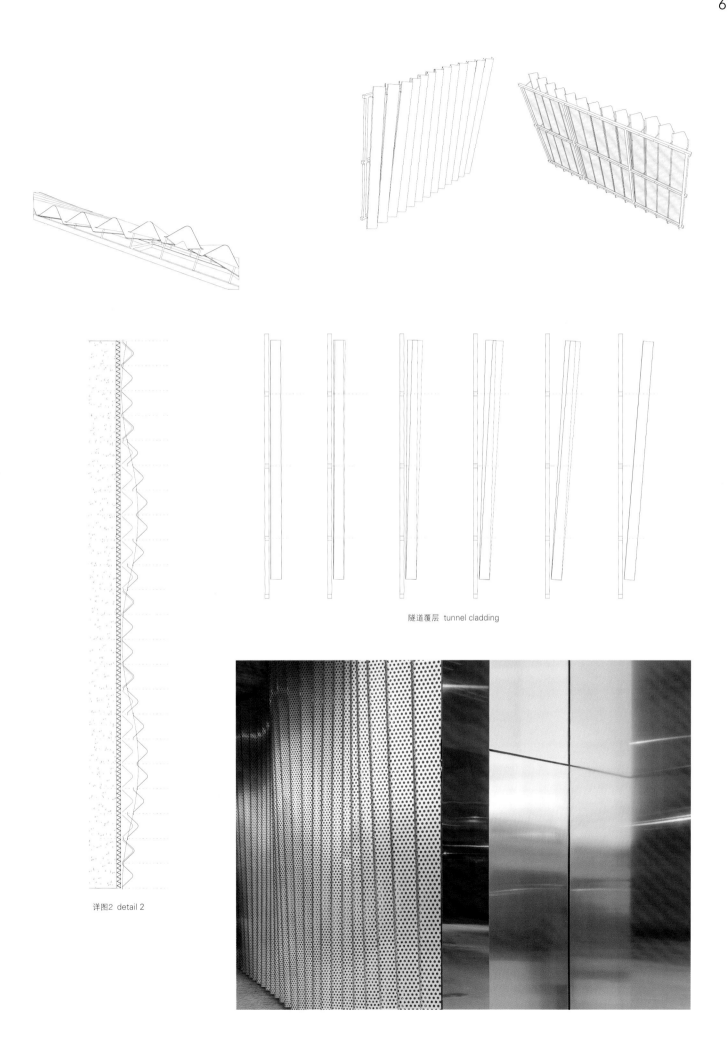

隧道覆层 tunnel cladding

详图2 detail 2

Nørreport 车站
Nørreport Station

Gottlieb Paludan Architects + COBE

哥本哈根市的Nørreport车站是丹麦哥本哈根市最繁忙的交通枢纽，每天能接纳约25万人次的旅客。它始建于1916年，在1934年进行了第一次现代化升级；2012年，为满足功能需求再次进行了基础性翻修改造。经过三年的建设，原来那个交通混乱的车站变成了一个开阔、舒适的城市空间，给行人和骑行者带来了方便。在项目设计初期，相关方就把有效提高交通流动性定为关键议题，同时，作为重中之重，提出要打造一个城市生活、休闲活动功能区，彰显哥本哈根大都市的活力与生机。基于以上出发点，设计者们决定不再把自行车停放处安排在不显眼的位置上，他们的想法是要彰显哥本哈根是世界上最佳的骑行圣地，而自行车停放代表了城市生活的重要一面。

项目建设分为三个部分：对长途列车站进行现代化升级；对支撑着隧道上方路面的桥梁结构进行改造；对站前广场及大楼、人行道和路面、自行车停放处、通道以及交通布局进行重新设计。Gottlieb Paludan建筑师事务所和COBE建筑事务所共同设计了这座新的车站和站前广场，以及这里所有的功能设施，该设计成功入围了2009年国际建筑设计竞赛。SWECO公司（以前的Grontmij公司）为该项目的工程顾问，而Bartenbach公司负责照明设计。

站前广场被设计成了城市"地板"的延伸部分。行人可以直接从周围的步行区走进车站，而过往车辆的线路经过重新布局，只留下一条站北交通干道。站前广场上设置了2100个自行车停车位，被戏称为"自行车床"，自行车摆放在下凹地带中，保证了清晰的层次，也能让人们毫无阻碍地看到整个空间。站前广场的几座建筑采用圆形造型，且主要为玻璃材质，给川流不息的人们提供了充足的移动空间，其透明的表面和自然流畅的造型也给人一种安全感。这个有凝聚力的空间没有背面，也没有死角。建筑和站前广场上自行车停放处的整体设计与布局都是根据对行人人流量的研究来设计的，他们有的来自周边马路，有的在这里穿过站前广场，有的是下楼梯后进入车站里面。这个项目中使用的材料包括白色混凝土、花岗岩、玻璃和不锈钢，之所以选择这些材料是因为它们拥有自然的表面，而且维护要求低。夜幕降临，这里的照明设施就会成为这里的一大特色，同时又是路标。而给地下站台通风的塔状建筑也一跃成为这个区域璀璨夺目的地标。

Originally established in 1916 and first modernised in 1934, Nørreport Station in Copenhagen is Denmark's busiest transport hub that serves about 250,000 train passengers and passers-by on a daily basis. In the need of fundamental renovation in 2012, three-year construction work has transformed the chaotic station into an open and comfortable urban space for pedestrians and cyclists. The efficiency of flow was a crucial aspect of the project proposal from the outset. In addition, priority was given in making a space for urban life and activities that reflects the vibrant, dynamic atmosphere of the metropolitan city of Copenhagen. In line

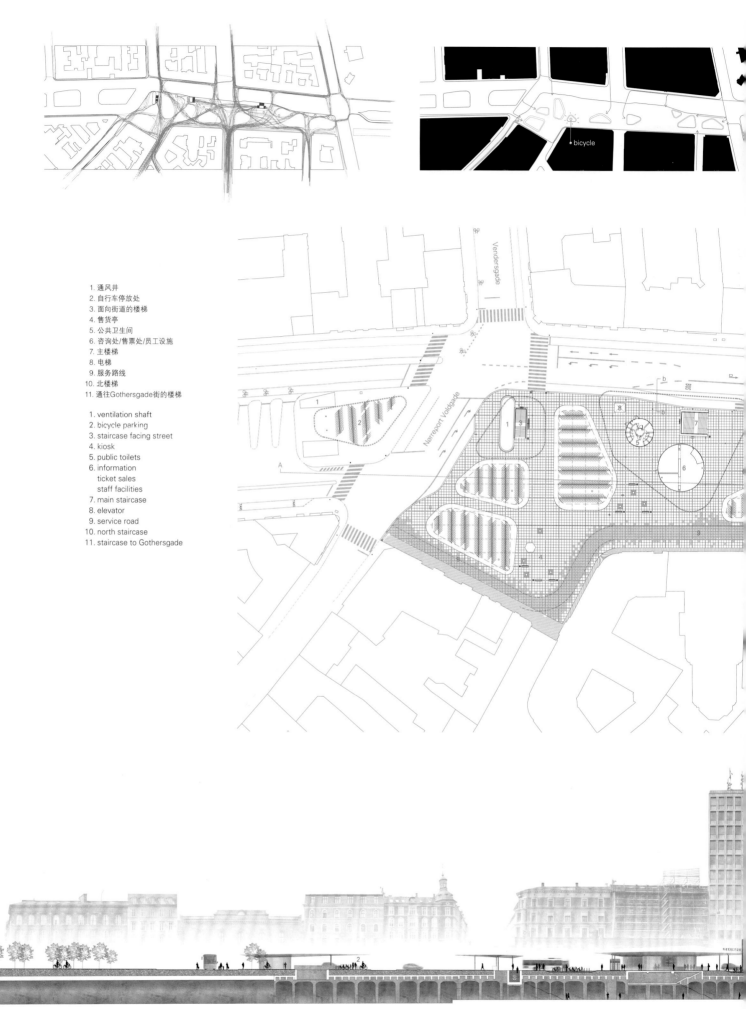

1. 通风井
2. 自行车停放处
3. 面向街道的楼梯
4. 售货亭
5. 公共卫生间
6. 咨询处/售票处/员工设施
7. 主楼梯
8. 电梯
9. 服务路线
10. 北楼梯
11. 通往Gothersgade街的楼梯

1. ventilation shaft
2. bicycle parking
3. staircase facing street
4. kiosk
5. public toilets
6. information
   ticket sales
   staff facilities
7. main staircase
8. elevator
9. service road
10. north staircase
11. staircase to Gothersgade

A-A' 剖面图 section A-A'

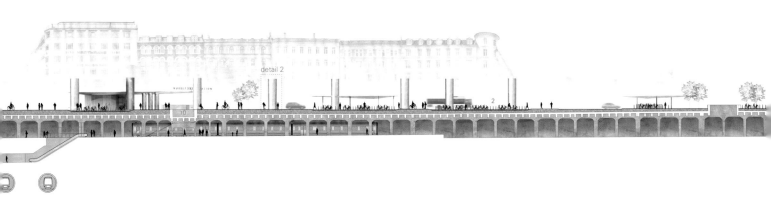

with this idea, the parked bicycles are not hidden but rather displayed as an important aspect of Copenhagen's identity as the world's best city for cyclists.

The project fell into three parts: modernization of the platform for long-distance trains, renovation of the bridge structures supporting the ground deck above the station tunnels, and the new designs of the station forecourt and buildings, pavings and surfacings, bicycle parking, access, and traffic arrangements. Gottlieb Paludan Architects and COBE completed the design of the new station and forecourt with all functions and facilities, having submitted as the winning entry in the international architectural competition in 2009. SWECO (previously Grontmij) was the engineering consultant and Bartenbach was in charge of lighting design.

The forecourt is designed as an extension of the city's "floor". Pedestrians can directly access the station from the surrounding pedestrianized zones, while vehicular traffic has been redirected, leaving only one traffic artery north of the station. Parking facilities are made for 2,100 bicycles as so-called "bicycle beds", which are recessed downwards in order to secure a clear hierarchy and unobstructed views of the whole space. The few buildings on the forecourt are built in rounded shapes, mainly using glass, providing enough room for the constant swarm of people. Its clarity and natural flow of the layout give people a sense of security. This cohesive space has no backs or corners. The overall design and layout of the buildings and bicycle parking facilities on the forecourt are based on a study of the flows of pedestrians from the surrounding roads and across the forecourt or down the stairs into the station. Materials used in the project – white concrete, granite, glass, and stainless steel, are chosen for natural surfaces and low maintenance demands. When darkness falls, the lighting becomes a feature as well as a means of navigation. The towers ventilating the underground platforms rise as luminous landmarks for the area.

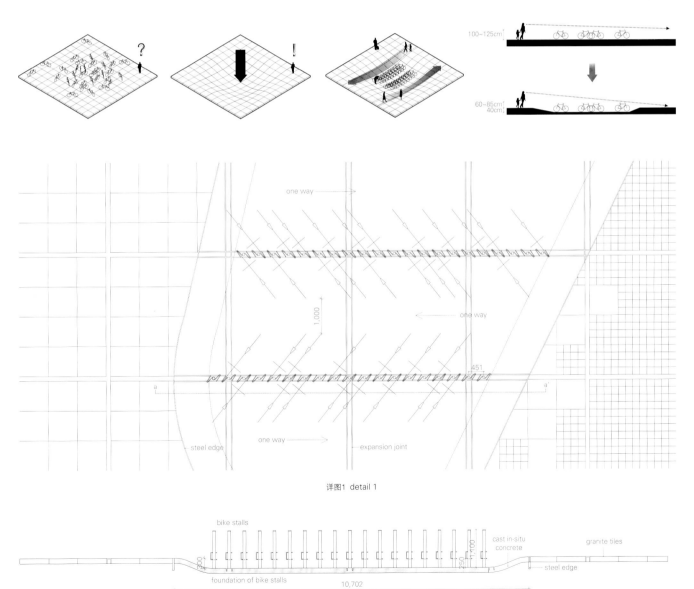

详图1 detail 1

a-a' 详图 detail a-a'

项目名称：Nørreport Station / 地点：Copenhagen, Denmark / 事务所：Gottlieb Paludan Architects, COBE / 项目团队：Marianne Jørgensens-lead architect; Ditte Holmgaard Griebel, Jens Peter Jørgensen-architect; Kenneth Bengtsson Hansen-architectual engineer; Preben Bang-constructing architect; Søren Gjerlev / 工程师&合作者：Grontmij (Sweco), Bartenbach Lichtlabor / 客户：Banedanmark (Rail Net Denmark), Danish State Railways (DSB), City Of Copenhagen / 公共空间面积：10,500m² / 自行车停车位数量：2500 / 造价：DKK 655 M / 税费：DKK 15 M / 竣工时间：2015 / 摄影师：courtesy of the Gottlieb Paludan Architects-p.68~69, p.71; ©Jens Lindhe (courtesy of the Gottlieb Paludan Architects)-p.70, p.72~73; ©Rasmus Hjortshøj-COAST (courtesy of the COBE)-p.62~63, p.66~67, p.75

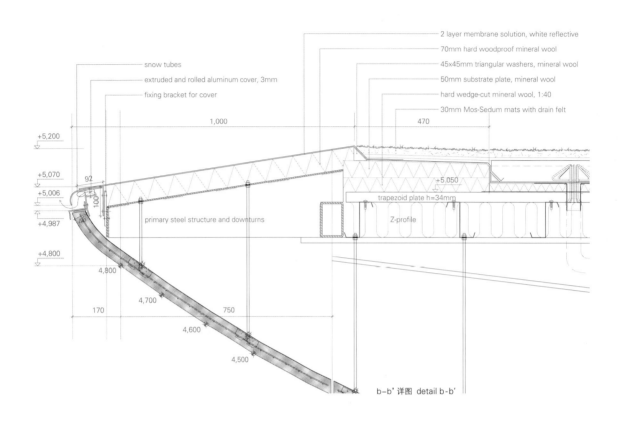

b-b' 详图 detail b-b'

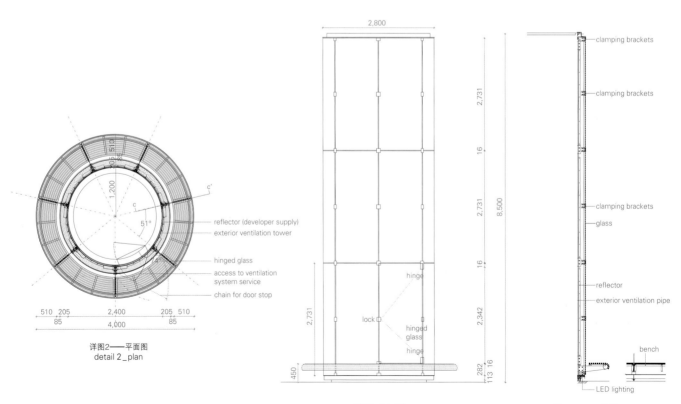

详图2——平面图
detail 2_plan

详图2 detail 2

详图2——c-c' 剖面图
detail 2_section c-c'

洛里昂布列塔尼火车南站
Lorient Bretagne South Railway Station

AREP

洛里昂多式联运枢纽项目是2017年"布列塔尼高速公路"工程的一部分，这条连通了坎佩尔、布雷斯特和巴黎三座城市的高速线路使人们可以用三个小时完成三个城市间的旅程。它内含不同类型的公共交通工具，包括铁路、城际公交和连接城郊地区的长途大巴。在南部地区重建的洛里昂布列塔尼火车南站靠近市中心，也是这个交通枢纽项目的中心。

总长超过115m的车站大楼呈线性造型，十分引人注目。大楼采用木质的巨型门式框架结构，暗示了作为城市传统产业的造船业。在洛里昂近年的建筑史上，一种突出的表现手法就是使用石头、混凝土或水洗混凝土等材料为主。与之呼应的是经过纤维增强的双层玻璃幕墙，它被用在了该项目不同区域（如车站入口、零售店、办公室）的所有洞口部分，能保护立面免受太阳直射。与凉廊一样，色彩也成为项目的一个背景元素。车站东立面呈敞开式面向城市，这一侧也是车站主入口所在的位置，欢迎着乘客的到来。南立面则由一个复杂的木结构体系组成，包括保温墙、室内外木覆层、双层玻璃表皮以及置于次级框架内的超高性能混凝土墙。与此形成对比的是，北立面大部分是玻璃幕墙，从其宽阔的洞口处，人们可以看到新建的快速区域线站台和铁路轨道，还能望见将在未来建成的北区通道以及Kerentrech的著名历史街区。整个北立面由大型玻璃板构成，玻璃板间嵌入跨距为4.8m的金属横梁。

车站大楼的结构是由花旗松木层压板制成的24个门式框架，跨度在12m至19m之间，高度为11m。建筑中央由混凝土核心筒支撑，而大厅和顶棚用混凝土门式框架支撑，这些框架所在区域的功能是保证建筑正常运营。车站大楼的屋顶还向外悬挑出一部分用作长途客运站的顶棚，悬挑的这部分由悬臂梁支撑，悬臂梁沿着顶棚南侧边缘延伸，呈弧形向下弯折后与地面相连。建筑师设计了一条城市步道，在车站开放期间，这条步道可以将乘客引导至站台，同时还起到将Kerentrech区与市中心连接起来的作用。建筑师还设计了一座行人天桥，横跨铁轨两侧，桥的长度是60m，每一半的跨度是30m。与车站主立面相连的公交站顶棚总长超过160m，支撑顶棚的梁每隔4.8m设置一根。站北广场和站南广场设有出租车候车室和落客区。另外停车场和自行车停放设施也即将纳入附近项目的建设方案中。

The Lorient multimodal hub is a part of the "Bretagne à Grande Vitesse" project (High-speed for Brittany) in 2017, a high-speed line which allows a three-hour total travel time among Quimper, Brest, and Paris. It accommodates different types of public transport means: rail, inter-city buses and coaches serving the conurbation. The Lorient-Bretagne South Railway Station is rebuilt on the south, close to the city center and in the heart of this transport hub.

The linear form of the building, stretching over 115m, gives the station an imposing presence. Along with the large timber portal frame forming the building's structure, it alludes to the city's shipbuilding tradition. Lorient's recent architec-

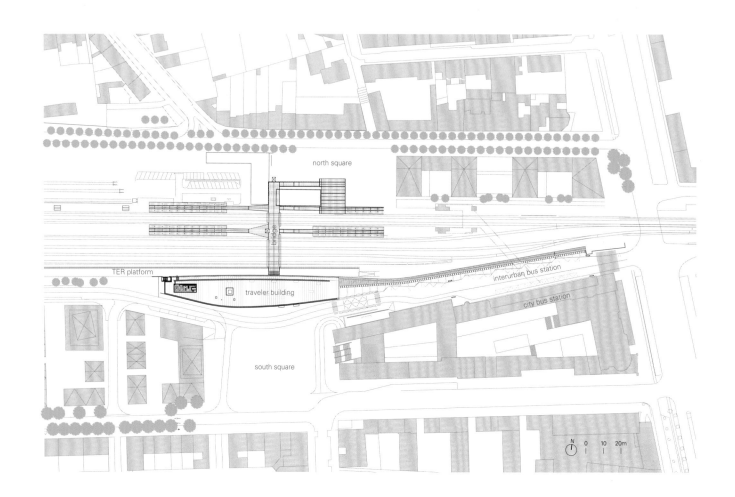

tural history gives prominence to stone as well as concrete and washed-concrete facades, all of which are echoed by the fibre-reinforced double skin featuring all the openings related to the various elements of the project (station entrance, retail outlets, offices) and protecting the facade from solar radiation. Similarly, color is a background element that resembles the loggias of Lorient. The east facade opens up to the city and welcomes the passengers as the main entrance to the station. The south facade is composed of a complex timber structure comprising insulation, interior and exterior timber cladding, double glazed skin, ultra-high-performance concrete (UHPC) screen modeled in sub-frames. In contrast, the north facade is mostly glazed and its large openings allow views on the new TER platform (express regional lines), the rail tracks, the future north access and the historic district of Kerentrech. It is gridded with large glass modules featuring metal crosspieces with a 4.8-m span between the beams.

The structure of the passenger building consists of 24 portal frames made of Douglas-fir laminated timber having a span of between 12m and 19m and a height of 11m. The central part of the building are braced by a concrete core while the hall and canopy are supported by concrete portal frames that are located in the area intended for operational purposes. The roof extends towards the coach station as a canopy, which is supported by a cantilevered beam running along its south edge and arching down to the ground. An urban walkway allows to access the platforms and links Kerentrech district to the city center during the station opening hours. The footbridge forms a 60m link between the two sides of the tracks and with a span of 30m each. The bus station shelter located by the main facade stretches out over 160m and forms a canopy supported by beams set every 4.8m. The north and south forecourts house taxi ranks and drop-off areas. Car parks and bike parking facilities will be incorporated in the neighboring construction projects later.

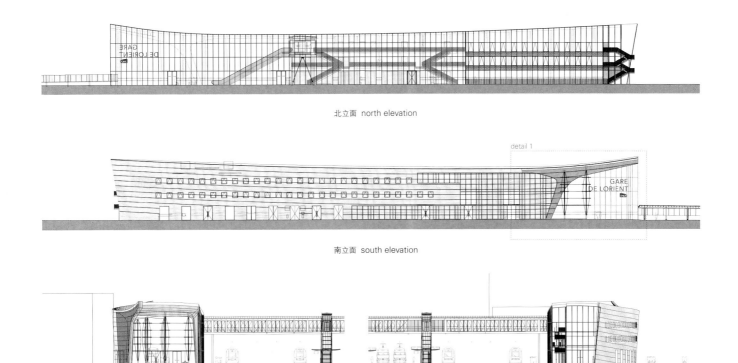

北立面 north elevation

南立面 south elevation

东立面 east elevation

西立面 west elevation

三层 second floor

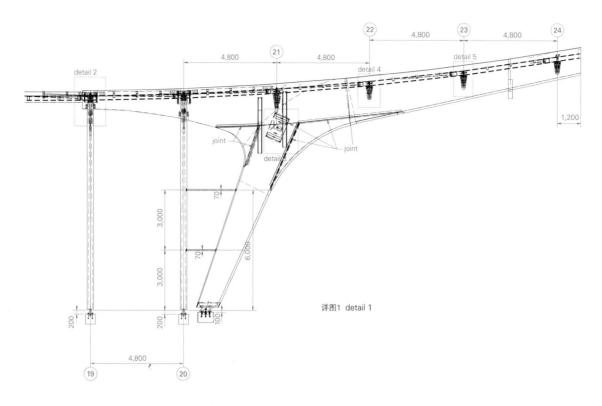

详图1 detail 1

项目名称：Lorient Bretagne South Railway Station
地点：Place Francois Mitterrand, 56100 Lorient, France
事务所、项目管理：AREP+SNCF Gares & Connexions
项目建筑师：Etienne Tricaud, Jean-Marie Duthilleul, François Bonnefille, Olivier Boissonnet
框架与立面工程顾问：H.D.A. Hugh Dutton & Associés, Mitsu
基础与混凝土结构：S.R.B.
外围护结构(木框架与金属结构)顾问：Mathis et Baudin Châteauneuf
立面设计：A.C.M.L.
客户：SNCF Gares & Connexions, SNCF Réseau, Lorient Conurbation
用地面积：6,000m² / 建筑面积：1,900m²
总楼面面积：2,320m²
造价：new passenger building - €9,750K; pedestrian bridge, intercity coach station and north access - €12,059K; offices - €2,390K, (tax excluded)
竣工时间：2017.5
摄影师：©D. Boy de la Tour (courtesy of the architect)

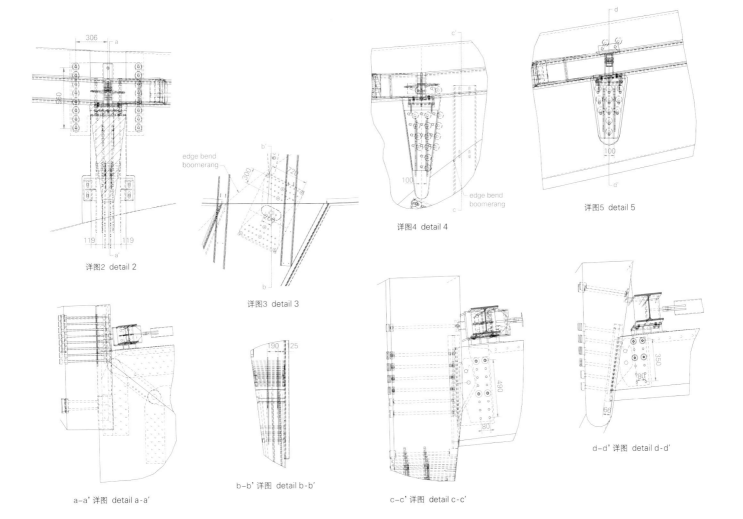

详图2 detail 2

详图3 detail 3

详图4 detail 4

详图5 detail 5

a-a' 详图 detail a-a'

b-b' 详图 detail b-b'

c-c' 详图 detail c-c'

d-d' 详图 detail d-d'

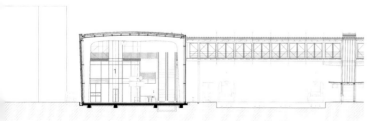

1. 客运大厅 1. passenger hall
A-A' 剖面图 section A-A'

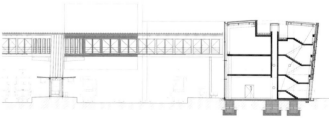

1. 车站外的办公室 1. offices out of station
B-B' 剖面图 section B-B'

1. 办公室和车站服务设施　1. offices and station services
C-C' 剖面图　section C-C'

1. 办公室和车站服务设施　1. offices and station services
D-D' 剖面图　section D-D'

那不勒斯阿夫拉戈拉高铁站
Napoli Afragola High Speed Train Station

Zaha Hadid Architects

那不勒斯阿夫拉戈拉火车站一期建设已经竣工，它即将成为意大利南部重要的交通枢纽站之一，同时为四条高速城际线、三条区间线路和一条当地通勤线路提供服务。这座火车站位于那不勒斯以北12km处，成为意大利南部铁路网的主要交会点，连接着意大利南部包含坎帕尼亚、普利亚、莫利塞、卡拉布里亚和西西里岛在内拥有1500万居民的地区，同时还连通意大利北部地区和欧洲其他国家和地区的铁路网络。它还负责把来自意大利北部以及欧洲其他国家和地区的货物和旅客运送到意大利南部的一些港口城市，例如，焦亚陶罗、塔兰托、巴里、布林迪西、巴勒莫和奥古斯塔等。在过去十年内，整个地区铁路运输的需求量增加了50%，并在未来会继续增加。鉴于此，政府在不断扩张的那不勒斯大都市东部新建的南北向铁路通道上建造了这座阿夫拉戈拉火车站。如果铁路全线投入运营，这个火车站每天的客流量将会达到32 700人，年客流量约1200万人，每天早晚高峰段时段通勤人数将达到4800人。

该车站的职能就像是一座城市公共连接桥，将铁路两侧的社区连接在了一起，根据旅客们来往的路线确定各进出站口的位置，缩短旅客步行距离，方便他们乘坐不同线路的列车。该项目还对八条铁路上面的公共人行道进行了扩建，把这条人行天桥变成了主要的客运大厅，其中设置了各种服务设施，为到达、出发和换乘的旅客提供服务，旅客可以直接从这里前往下面的所有站台。

旅客来往路线也决定了站内各空间的几何形状。车站两端的大型入口可以迎接和指引旅客去往高层公共区域，里面设有商店等服务设施。从车站两边进来的旅客在中央大厅会合，从这里往上走就是咖啡馆和餐厅。这个建在铁路轨道上方的中央大厅成为阿夫拉戈拉一个全新而必要的公共空间，旅客们可以由此去往下面的各个站台。

该项目于2015年投建，使用钢筋混凝土地基支撑上面长达450m的大厅。大厅采用玻璃屋顶，其走向呈弯曲的不规则形状，因此用到了200根不同形状的钢肋板，肋板上覆盖可丽耐材料。车站内采用的混凝土构成成分特殊，能把弧形混凝土构件的性能发挥到极致。制作混凝土的模板不再使用木材，而是采用预制钢板，通过CNC铣削聚苯乙烯模板实现其双曲的形状。车站的中央大厅设计旨在让建筑体现出生态可持续的特性。屋顶的集成太阳能电池板，通过结合自然光和通风设备以及地源冷却/加热系统为整座建筑提供能源，让车站能够最大限度地降低能耗。

任何伟大的国家都需要在伟大工程建设上取得飞跃式的进步。阿夫拉戈拉新车站是促进南部经济发展的基础设施项目的基石。多式联运中心的落成让人们看到这个国家的成长和进步，也展示出艺术和工程技术完美结合后的产物。

The first phase of Napoli Afragola Station, one of southern Italy's key interchange stations serving four high-speed inter-city lines, three inter-regional lines, and a local commuter line, has been completed. Located 12km north of Naples at the major intersection, Napoli Afragola connects the 15 million residents of Campania, Puglia, Molise, Calabria, and Sicily in southern Italy with the national rail network in the north and the rest of Europe. It also enables goods and passengers from Europe and northern Italy to access the southern ports of Gioia Tauro, Taranto, Bari, Brindisi, Palermo and Augusta. To meet future demand for rail travel throughout the region (which has increased by 50% in the past decade), the Napoli Afragola station is within the new north/south rail corridor in the east of the greater Napoli

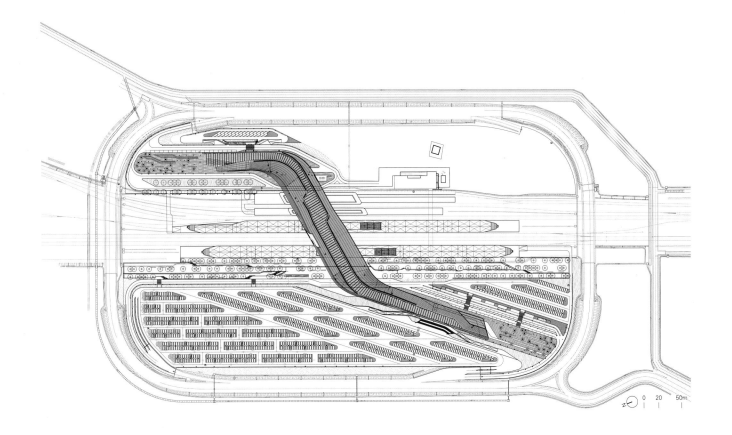

项目名称：Napoli Afragola High Speed Train Station / 地点：Napoli Afragola, Italy / 事务所：Zaha Hadid Architects / 设计：Zaha Hadid, Patrik Schumacher
项目指导：Filippo Innocenti / 项目合伙人：Roberto Vangeli / 项目建筑师（竞赛阶段）：Filippo Innocenti, Paola Cattarin / 现场监理团队：Marco Guardincerri, Michele Salvi, Pasquale Miele (BC, Building Consulting) / 设计团队：Michele Salvi, Federico Bistolfi, Cesare Griffa, Mario Mattia, Tobais Hegemann, Chiara Baccarini, Alessandra Bellia, Serena Pietrantonj, Roberto Cavallaro, Karim Muallem, Luciano Letteriello, Domenico Di Francesco, Marco Guardincerri, Davide Del Giudice, Paolo Zilli
竞赛团队：Fernando Perez Vera, Ergian Alberg, Hon Kong Chee, Cesare Griffa, Karim Muallem, Steven Hatzellis, Thomas Vietzke, Jens Borstelmann, Elena Perez, Robert Neumayr, Adriano De Gioannis, Simon Kim, Selim Mimita / 结构工程与土工技术：Hanif Kara, Paul Scott-Akt; Giampiero Martuscelli
- Interprogetti Environmental engineering, M&E: Henry Luker, Neil Smith-Max Fordham; Francesco Reale, Vittorio Criscuolo Gaito-Studio Reale
建筑规范、当地协调团队：Alessandro Gubitosi - Interplan 2 Srl Costing: Pasquale Miele - Building Consulting / 防火：Roberto Macchiaroli - Macchiaroli & Partners Srl
景观设计：Eelco Hooftman-Gross Max / 交通工程：Max Matteis-JMP / 声学设计：Paul Guilleron-Paul Guilleron Acustics / 构造设计：Prof. Ing. F. Sylos Labini, Ing D. Sylos Labini-Sair-Geie; Rocca Bacci Associati / 承包商：Astaldi S.P.A., NBI S.p.A-Ati Astaldi S.P.A. / 客户：Rete Ferroviaria Italiana S.P.A. / 用途：train station, retail, hospitality / 用地面积：190,000m² / 总楼面积：30,000m² / 零售区与酒店面积：10,000m² / 材料：steel, metal carpentry, cladding panel, glass façade cladding
竞赛获胜时间：2003 / 施工开始时间：2015.5 / 竣工时间：2017.6 (Phase 1) / 摄影师：©Hufton+Crow (courtesy of the architect)

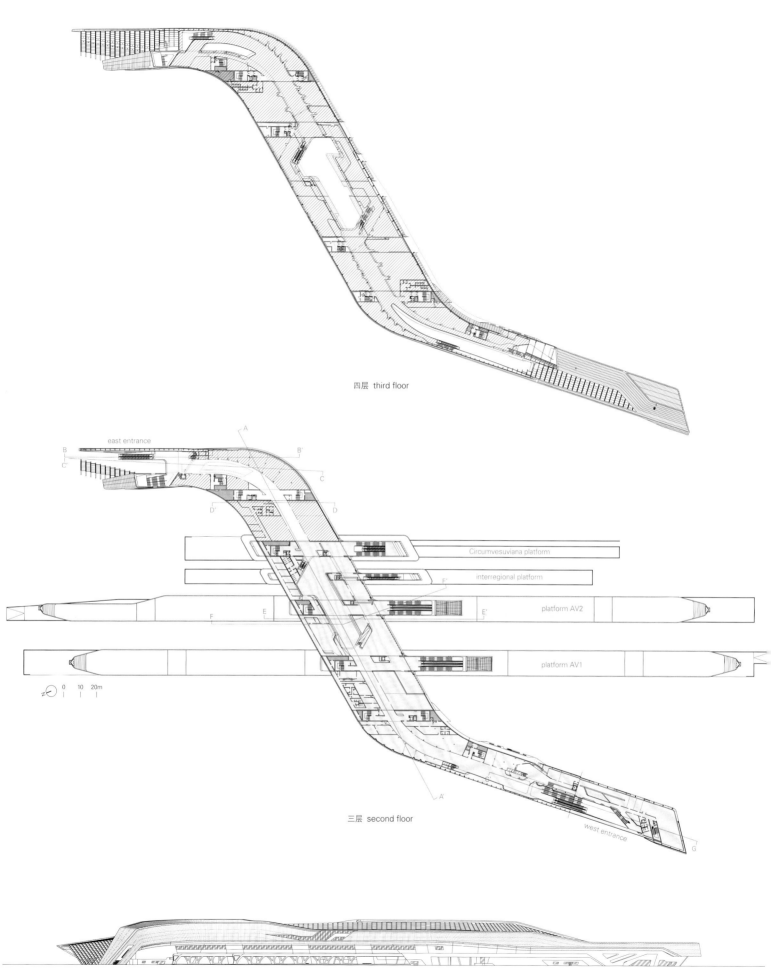

metropolitan area. Once all lines operate, 32,700 passengers are expected to use the station every day (approximately 12 million passengers each year) with 4,800 commuters using the station each morning and evening rush-hour. Designed as an urbanised public bridge connecting the communities on either side of the railway, the station is largely defined by the circulation routes of passengers, minimising distances for those embarking and alighting, as well as connecting the passengers to the different train services. The station enlarges the public walkway over the eight railway tracks to such a degree that this walkway becomes its main passenger concourse – a bridge housing all the services and facilities for departing, arriving and interchanging passengers, with direct access to all platforms below. The paths of passengers have also determined the geometry of the spaces within. Large entrances at both ends of the station welcome and guide visitors up to the elevated public zones lined with shops and other amenities. Visitors from either side of the station meet in a central atrium overlooked by cafes and restaurants. This central atrium above the railway tracks is a much-needed new public space for Afragola and the main concourse where rail passengers

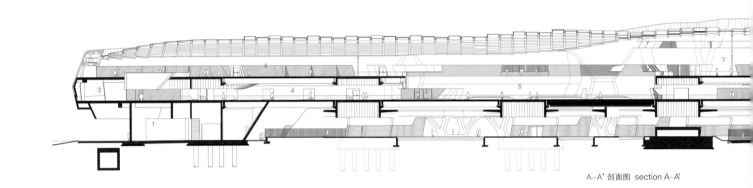

A-A' 剖面图 section A-A'

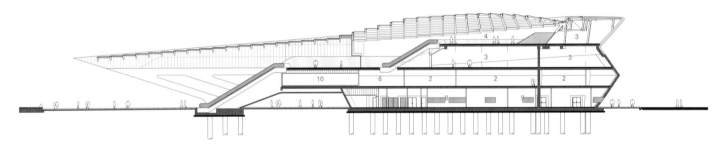

B-B' 剖面图 section B-B'

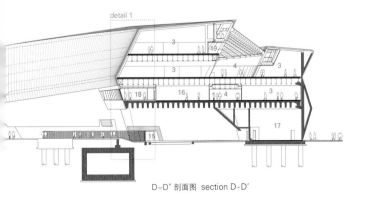

D-D' 剖面图 section D-D'

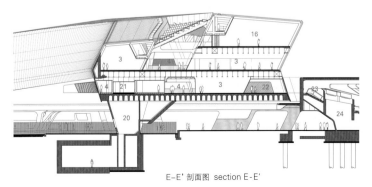

E-E' 剖面图 section E-E'

descend to the platforms.

First began in 2015, the station has been constructed as a reinforced concrete base supporting the elevated concourse, an extrusion of a trapezoid along a 450m curved path, that is made of 200 differently-shaped steel ribs clad in Corian®, with a glazed roof. The concrete used for the construction is a specific composition that provides optimum performance for curved structural concrete elements. Its wooden formwork is replaced by prefabricated steel units, and double-curves are realised with formwork created from CNC milled polystyrene models. The main concourse of the station is oriented towards ecological sustainability. Integrated solar panels in the roof, combined with natural light and ventilation as well as ground source cooling/heating systems will minimise energy consumption.

Any great country needs great projects that are a leap forward. The new station at Afragola is the foundation of the infrastructure program that promotes economic development in the south. The inauguration of the intermodal hub sends a message that the country is growing, symbolizing a wonderful combination of artistic and engineering expertise.

1. 货物装卸区	1. loading / unloading of goods
2. 储存区	2. storage area
3. 商业区	3. commercial area
4. 连接区	4. connective
5. 中央大厅	5. central atrium
6. 维护区	6. maintenance
7. 隧道	7. tunnel
8. 发电间	8. power plant
9. 当地UTA	9. local UTA
10. 维护办公室	10. maintenance office
11. IMP冷藏室	11. IMP refrigeration
12. UPS连续区	12. UPS continuity
13. 锅炉房	13. boiler room
14. UFF维护区	14. UFF maintenance
15. 前门廊	15. front porch
16. 餐厅	16. restaurants
17. 停车场维护区	17. park maintenance
18. 档案室	18. archive
19. 东隧道	19. east tunnel
20. Cavedio区	20. Cavedio
21. 卫生间	21. toilet
22. VIP中心	22. VIP center
23. 通往码头的楼梯	23. staircase to the pier
24. 技术间	24. technical room
25. 高速铁路线站台	25. high speed line platform
26. 售票处	26. ticket office
27. 区间铁路线站台	27. interregional line platform
28. 车辆租赁区	28. car rental
29. 等候室	29. waiting room
30. 安全梯	30. security ladder
31. 通往站台的楼梯	31. access to the platform
32. 西侧隧道	32. west tunnel

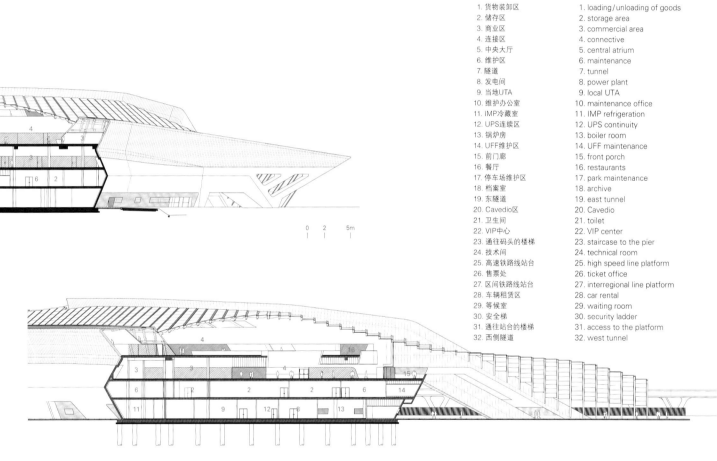

C-C' 剖面图 section C-C'

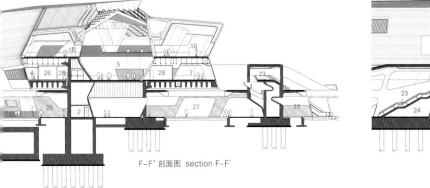

F-F' 剖面图 section F-F'

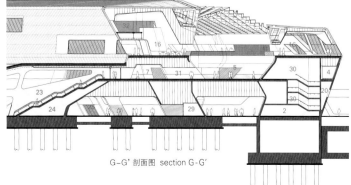

G-G' 剖面图 section G-G'

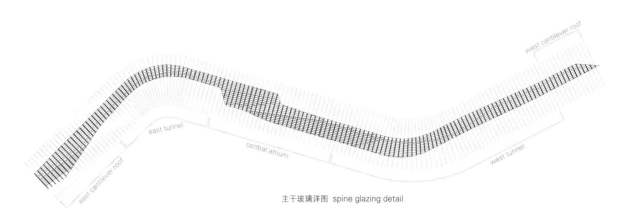

主干玻璃详图 spine glazing detail

立面几何形状 facade geometry

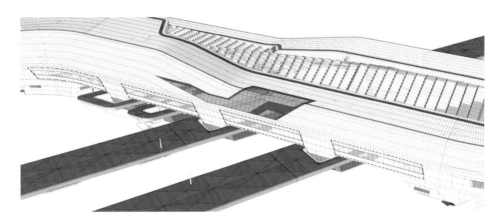

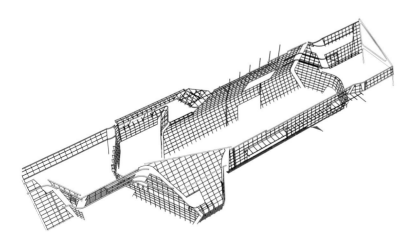
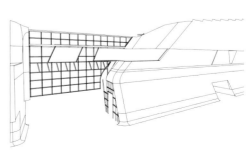

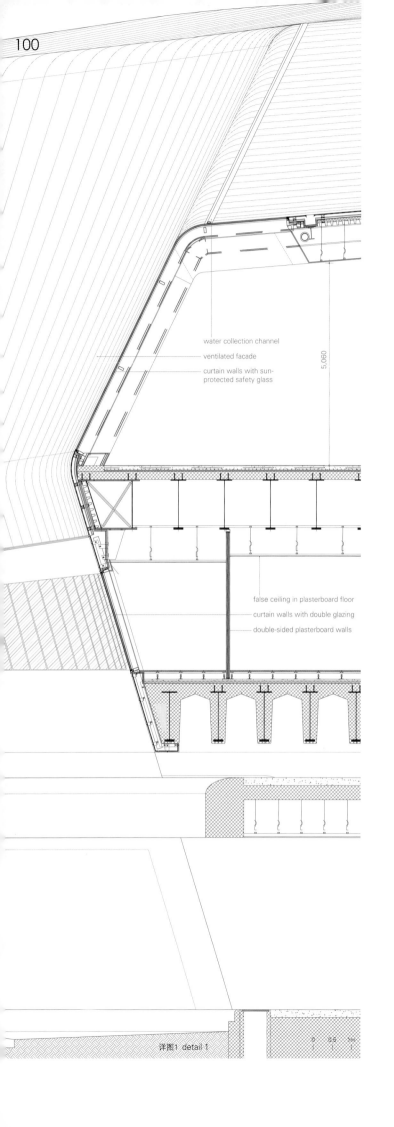

water collection channel
ventilated facade
curtain walls with sun-protected safety glass

false ceiling in plasterboard floor
curtain walls with double glazing
double-sided plasterboard walls

详图1 detail 1

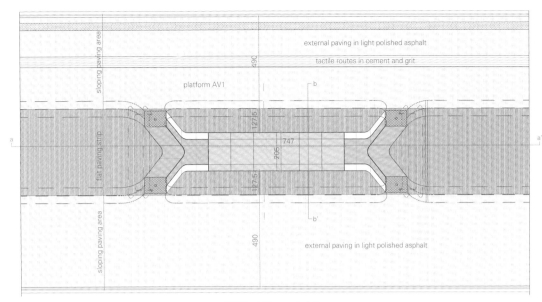

高铁站台 high speed platform

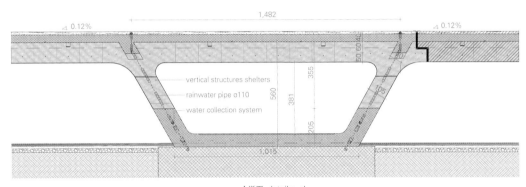

a-a' 详图 detail a-a'

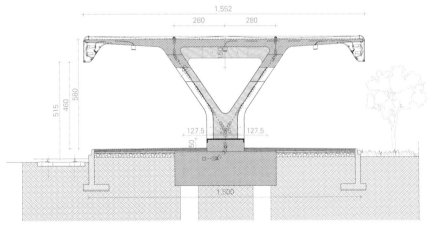

b-b' 详图 detail b-b'

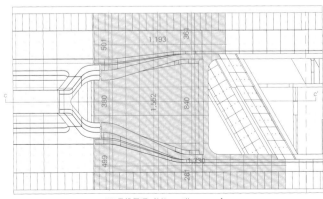

AV1悬挑屋顶 AV1 cantilever roof

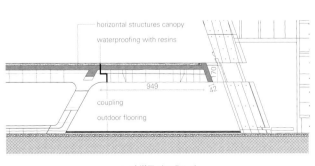

c-c' 详图 detail c-c'

香港西九龙站
Hong Kong West Kowloon Station

Andrew Bromberg at Aedas

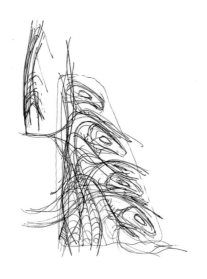

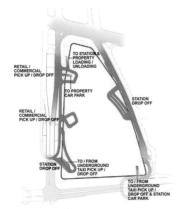

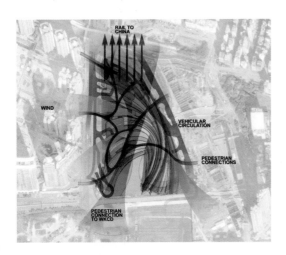

广深港高速铁路香港段长142km,连接国家高速铁路网,一直通到北京,全长超过25 000km。位于香港市区中心地带的广深港高铁站,占地430 000m²,设有15条轨道,将成为世界上最大的地铁终点站。车站本身还有一个隐含的目标,那就是车站作为通往香港的门户,必须要与周围环境相契合,仿佛在告诉旅客们"这里是香港",是他们即将走进或离开的地方。

香港西九龙站坐落于维多利亚港旁,紧邻未来的西九龙文化区,其设计理念应充分响应民众的要求。为了实现这一点,建筑师最大限度地减少了支撑空间,为下方的出发大厅留有充足的上空空间。外面的地面层朝大厅方向向下弯曲,而上方的屋顶结构则指向维多利亚港,由此打造出了一个45m高的空间。透过建筑的南立面,可以欣赏到香港中环的天际线和远处的太平山山顶。

项目设计的灵感来自香港这座城市的汇聚力,正如来自四面八方的轨道汇聚到这个车站一样。建筑师在建筑的内部和外部都试图最大化地展现百姓的姿态。

人行道几乎向上延伸到车站的整个屋顶,屋顶距离地面25m,上面是茂密的植物雕塑花园,园中景观一直延伸到下方的绿地,最终形成35 000m²的开放空间,在这个公共空间可以将维多利亚港和香港天际线的美景尽收眼底。

主厅的内部几乎像一片森林，倾斜的钢柱有力地支撑着屋顶，4000块玻璃板组成的幕墙将自然光引入建筑内部，使旅客即使从车站较低的位置也能看到这座城市。

香港的建筑大多都是垂直而密集地矗立在地面上，但这个车站的设计则显得与众不同，它俯卧在地面上，与周围的景观融为一体。建筑师安德鲁·布隆伯格是凯达建筑设计公司的建筑师，他热衷于让建筑具有如同液体一般的流动性，他的设计仿佛使来自各地的力量都汇聚到这个国际化都市中来。

布隆伯格说："我已经在这个项目上付出了近10年的努力，如今车站里人流穿梭的景象令我感到十分有成就感。新车站让人们有机会去寻找与香港这座城市之间的新联系。"

The new section of the Guangzhou-Shenzhen-Hong Kong high-speed rail service, which is 142km in length, connects with the National High Speed Rail network all the way to Beijing, the total length of which is more than 25,000km. Located centrally in Hong Kong, within the city's existing urban realm, the 430,000m² facility with fifteen tracks will be the largest below-ground station terminus in the world. Within the station itself, there was one underlying goal of the scheme. Acting as the "gateway" to Hong Kong, it was considered vital to connect the station with the surrounding urban context and make travelers aware of their arrival or departure, announcing: "You are in Hong Kong."

The site's prominence immediately adjacent to the future West Kowloon Cultural District and next to Victoria Harbour required a design that was completely motivated by civic demand. To achieve this, the design compacted all of the supporting spaces more efficiently to allow for a very large void down into the departure hall below. The outside ground plane bends down towards the hall, and the roof structure above gestures towards the harbour. The result is a 45m high volume which focuses all attention through the south facade towards views of the Hong Kong Central skyline and Victoria Peak beyond.

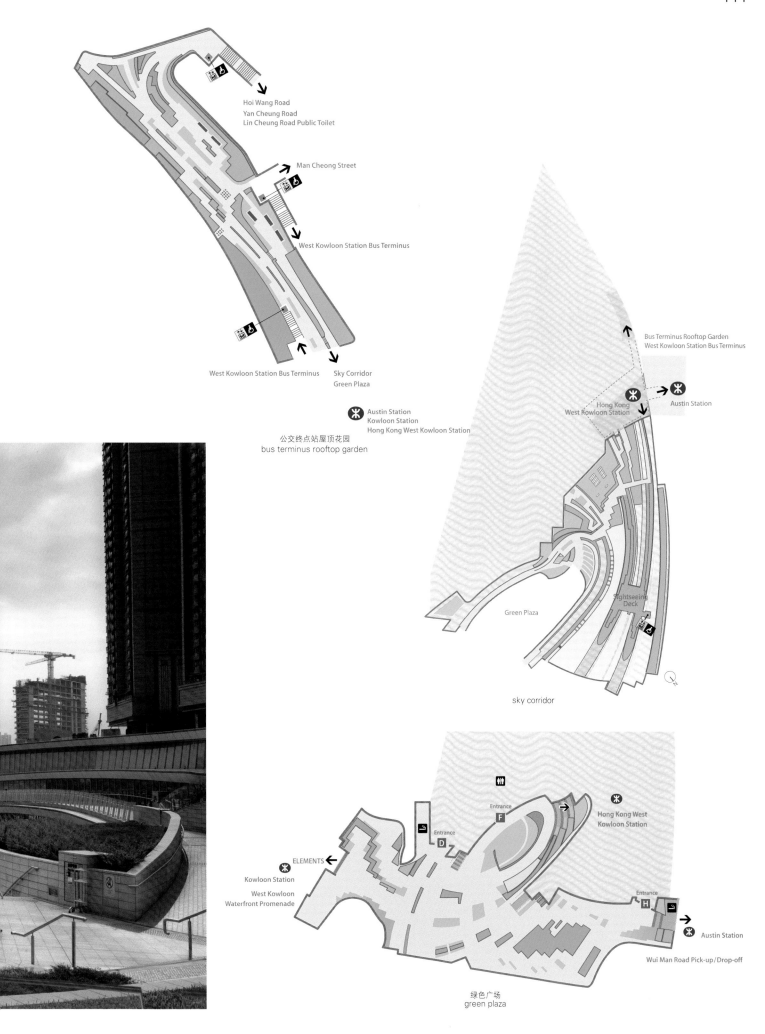

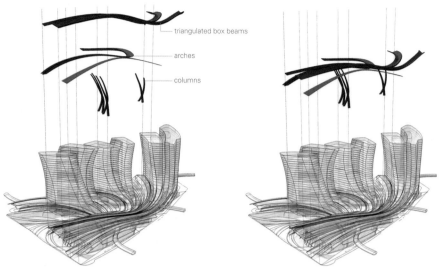
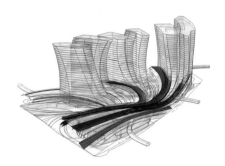

屋顶工程
roof engineering

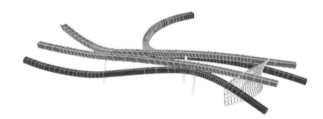

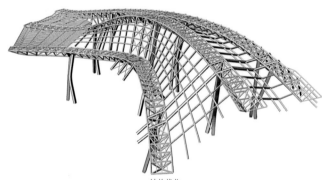

结构优化
structural optimizatioin

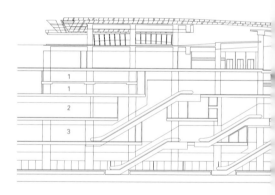

项目名称：Hong Kong West Kowloon Station
地点：West Kowloon, Hong Kong
事务所：Andrew Bromberg at Aedas
项目经理：AECOM
细节设计顾问：AECOM - Aedas Joint Venture
结构工程师：AECOM, Buro Happold
机电顾问：Meinhardt / 立面顾问：ALT
景观建筑师：EDAW / 工料测量师：Windell
总承包商：Leighton – Gammon Joint Venture
其他参与方：FMS, ROSTEK, MVA, Atelier Pacific
客户：MTR Corporation Hong Kong
用地面积：58,797m² / 总楼面面积：430,000m²
开放空间面积：over 30,000m²
建筑高度：29m / 竣工时间：2018
摄影师：
©Paul Warchol (courtesy of the architect) - p.104~105, p.108~109, p.110 bottom, p.118
©Virgile Simon Bertrand (courtesy of the architect) - p.106~107, p.110 top, p.112~113, p.116~117, p.120~121

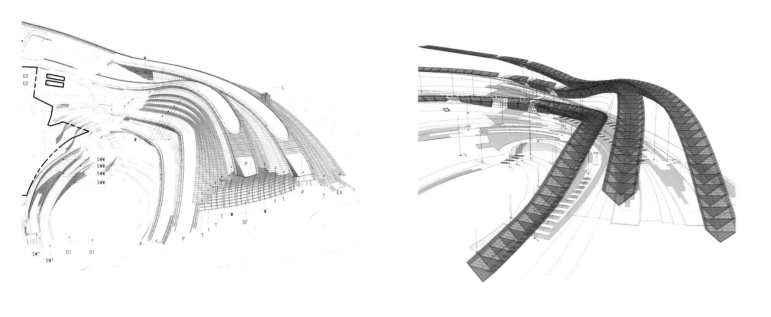

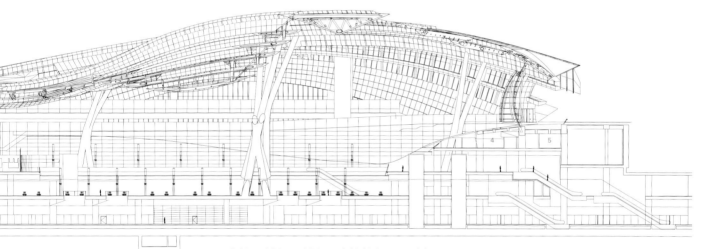

1. 停车场 2. 到达大厅 3. 出发大厅 4. 车站机电设备区 5. EAA走廊
1. property car park 2. arrival CIQ 3. departure CIQ 4. M/E zone for STA 5. EAA corridor
纵剖面 longitudinal section

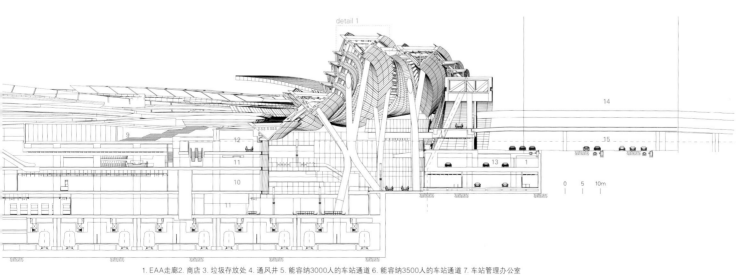

1. EAA走廊 2. 商店 3. 垃圾存放处 4. 通风井 5. 能容纳3000人的车站通道 6. 能容纳3500人的车站通道 7. 车站管理办公室
8. 卫生间 9. 广场 10. 商务休息室 11. 车站 12. VIP室 13. 出租车排队点 14. 通往奥斯汀车站的入口 15. D1路
1. EAA corridor 2. store 3. refuse store 4. vent shaft 5. 3,000 clear for station passage 6. 3,500 clear for station passage
7. STA management office 8. toilet 9. plaza 10. business lounge 11. STA 12. VIP 13. taxi queueing 14. assess to Austin station 15. road D1
横剖面 cross section

The organization of the design was inspired by the idea of forces converging on Hong Kong – likened to the converging tracks coming into the station. The project maximizes civic gestures both internally and externally.

The pedestrian paths flow up and access almost the entire rooftop of the station itself, 25m above ground, in a densely vegetated sculpture garden and landscaped extension of the green below. The resulting 35,000m² open space offers a spectacular public vista over Victoria Harbour and towards Hong Kong's skyline.

The interior of the main hall is almost like a forest, with leaning steel columns robustly supporting the rooftop and curtain walls laid with 4,000 glass panels to bring natural daylight into the building and a glimpse of the city, even from the lower levels of the station.

Unusual for the dense vertical city of Hong Kong, the design of the station closely hugs the ground, merging with the surrounding landscape. The architect, Andrew Bromberg at Aedas, was keen to bring in the sense of fluidity, reflecting various forces converging in this global city.

"I have been working on this station for nearly 10 years and it's great to see it now teeming with people. The new station is an opportunity for all people to discover new connections to the city of Hong Kong," says Andrew Bromberg.

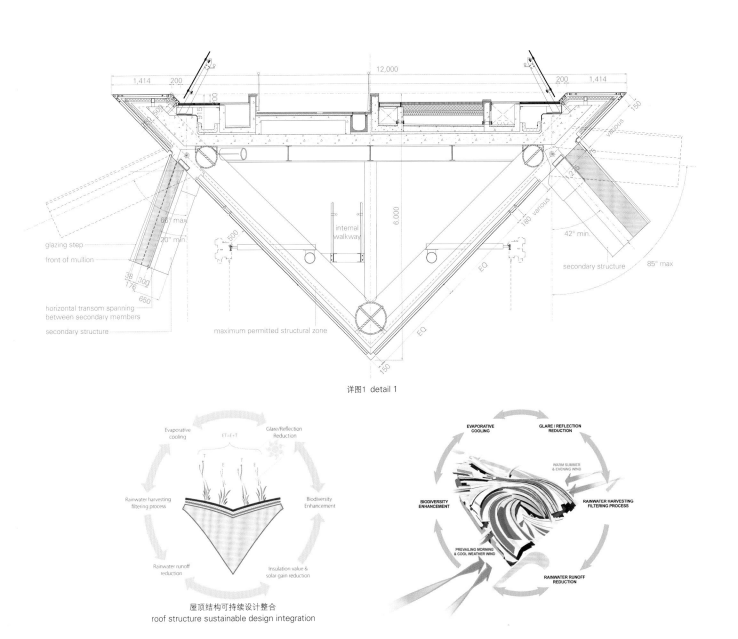

详图1 detail 1

屋顶结构可持续设计整合
roof structure sustainable design integration

城市基础设施走向光明的未来

Toward a B
of Urban In

过去很长一段时间里，基础设施的建设给城市留下了种种"伤痕"，其存在的唯一原因就是为达到各种功利的目的。正因如此，除了机场航站楼之外，我们城市基础设施的建筑美学都来自于地堡、坟墓、厂棚等各种严严实实的包围式结构，既不采光也不向外透光。现在，建筑师们对城市建设已经有了更全面的认识，他们决心一定不再重蹈覆辙，而是积极地寻找新的方法，使光这个元素彻底改变城市基础设施的魅力。从基本的公交车站，到精致的海滨码头，从温暖的火车站，再到安全的停车场，建筑师们开始利用光线将这些人们不想要却又不可或缺的

For too long, urban infrastructure blighted our cities, as their utilitarian purposes monopolized their reason for existing. As a result, with the exception of airport terminals, our urban infrastructure borrowed from the aesthetics of bunkers, tombs, sheds, and any enclosure in which light was not permitted in, nor emitted out. Now, with a more comprehensive understanding of cities, and a solid determination not to repeat the mistakes of the past, architects are finding new ways to revolutionize the appeal of urban infrastructure with light. From basic bus stations, to elaborate seaside terminals, and from warm train stations, to safe parking complexes, architects are using light to turn

ight Future
rastructure

设施变成迷人的城市建筑作品。本文根据功能的不同,细分了15座城市建筑,向读者展示建筑师采取的种种做法来灵活运用光线,给建筑平添魅力,这些做法有共性部分,也有独到之处。有的把光用作保暖和保证安全的工具;有的利用光来活跃气氛或者是给空间注入能量;还有的通过结构的轻盈来增强光感。每一个案例都把光视为提升基础设施魅力的必要元素,并附有建筑师对项目期望的描述。

these unwanted but necessary structures, into appealing works of civic architecture. In this article, 15 urban architectural projects are subdivided based on function, looking at how each project is enhanced by light. Similarities and differences are found across all projects in how each benefits from more light. In certain projects, light is used as a tool for warmth and safety. In one instance, light is used to enliven spirits, and in another, used to infuse a space with energy. In other projects, light is enhanced by the lightness of structure. In every case, light is used to raise the appeal of urban infrastructure, with the architects describing the desired outcomes.

Toward a Bright Future of Urban Infrastructure

Phil Roberts

　　设计多式联运枢纽站的最大难题在于要把多种交通方式拼接成一个有凝聚力的整体。根据法律规定，高速列车、城际公交车、长途大巴和小型地面交通工具都有不同的减速、停车距离，以保证乘客的安全。而一旦像这样把着眼点放在设施的功能性上，往往就会使整个工程效果显得笨重拖沓。对于AREP建筑事务所的Oliver Boissonnet来说，设计洛里昂南布列塔尼火车站（76页）时面临的最大难题就是要想办法减少项目因其复杂性而给乘客们带来的压迫感。他解释道："我要设计出一组轻盈细长的木质支撑结构，给乘客打造一个清晰可辨、安全友好的车站，这对我来说是一个极大的挑战。"客运大厅高耸的木结构框架要比钢结构框架更有温度。每一个木框架旁边的地板照明设备更是增强了这种平易近人的品质，这些设备产生的光晕温暖了周围少数金属构件冰冷的外观。这种材料之间的温度变化在太阳升起时效果最明显，框架的发光效果会比夜晚更加显著。

　　哥本哈根Nørreport车站（62页）是一个交通枢纽，提供了自行车停放处，并将它们分成四个区域，这四个区域朝向四个街区延伸，形成四个轻微下凹的停车"口袋"。从地面上看，整个交通枢纽从西南向东北延伸，但在地下，地铁则沿着东南至西北的轴线行驶。依照这个布局，Nørreport车站被放置在了十字交叉点上，成为连接北侧繁华大道和南侧行人广场的纽带。每个自行车停车"口袋"都能为工作了一天的人们减轻从2100个停车位中寻找自己车辆的烦恼。从通风塔、数百个单独的护柱和落地灯发出的光，投射到车站顶棚的下面，在晚上会形成令人安心的光芒。Gottlieb Paludan建筑师事务所和COBE建筑事务所共同发表了项目说明，文

One challenge with multimodal hubs is stitching together many modes of transportation into a cohesive whole. By law, high-speed trains, inter-city buses, coaches, and smaller ground vehicles, each require different distances for slowing down, for stopping, and passenger safety. A focus on functionality can leave such projects seeming too heavy. For AREP's Oliver Boissonnet, the critical issue to overcome for the Lorient Bretagne South Railway Station (p.76) was making such a complex project less intimidating for passengers. "The major challenge was to design a lightweight and slender timber structure that serves a legible, safe and welcoming station," explains Boissonnet. The towering timber frames of the main hall of the hub are much warmer in wood than they would have been with steel. The floor lighting next to every wooden frame enhances this approachable quality, creating a shimmer that redeems the cold appearance of the few metallic objects around. This transfer of warmth between materials is even more apparent when the light of sunrise makes the frames glow even more pronounced than at night.

Nørreport Station (p.62), a transit hub in Copenhagen, provides bicycle parking and splits them up into slightly sunken pockets, stretched out over four blocks. On the surface, the hub spans from southwest to northeast, but below grade the subway trains roll in a southeast-northwest axis. This configuration makes Nørreport a cruciform hub which acts as the bond that unifies the bustling thoroughfare to the north, and a pedestrian plaza to the south. Each pocket simplifies the task of finding a bicycle among the 2,100 spots after a long day of work. Light emitted from the ventilation towers, the hundreds of individual bollards, and floor lights, reflects on the underside of the canopies, creating a reassuring glare at night. According to the joint text by Gottlieb Paludan Architects and COBE: "The overall design and layout of the buildings and bicycle parking facilities

Nørreport 车站，丹麦
Nørreport Station, Denmark

中这样说道："建筑和站前广场上自行车停放处的整体设计与布局都是根据行人的人流量来设计的，他们有的来自周边马路，有的在这里穿过站前广场，有的是下楼梯后进入车站里面。"即使是在日落之后，旅客们也可以轻松地在车站空间中穿行。由于各个方向都会有旅客走来，所以设计师选用多样化的照明方式，这样便可以消除空间中的阴影区域，让旅客知道别人能看见自己，从而感到安心。

多伦多的先锋村站（26页）采用了奇思妙想的风格和光线设计，让通勤变得不再那么难以忍受。据已故的威尔·阿尔索普说，加拿大的冬季最能彰显出先锋村的魅力。"我不得不假设，在一个个十分寒冷、痛苦的一月清晨……车站里有一部分人又得开始为接下来一堆不愿意做的工作而奔波。我想让他们感受到一丝喜悦，体会一丝快乐。"进入像教堂一样的主大厅后，通勤者们就会急匆匆地下楼去赶地铁，但有时会停下来看看好玩的景象。

在西雅图的盎格鲁湖车站和广场的主体部分，可以明显地看到舞台效果成为引入光线的一种方法（132页）。停车场的结构隐匿在富有动感的阳极氧化铝板下，勾勒出混凝土板波浪起伏的轮廓。整个项目的灵感来源于威廉·弗西斯的舞蹈作品《舞蹈几何》。阳光穿过结构外部这条"舞裙"，形成一条条光带打在地板上。停车场和车站由一条连廊衔接，上面是弯曲的广场顶棚，将整个项目的不对称性与轻轨车站联系在一起。设计师想尽一切办法使平常严肃的实用建筑具有魅力。主设计师劳伦斯·斯卡帕

on the forecourt are based on a study of the flows of pedestrians from the surrounding roads and across the forecourt or down the stairs into the station." Users feel at ease moving through the space, even after sunset. Since the flow of users is multidirectional, and the diversity of lighting options removes shadowy zones within the space, the knowledge of always being seen is comforting.
Toronto's Pioneer Village Station (p.26) uses a whimsical style and light to make the commute more bearable. According to the late Will Alsop, Pioneer Village's charm will be most effective during Canadian winters. "I have to assume that on one of those really cold, miserable January mornings… there will be people down there travelling to a job that they'd probably rather not do. What I want them to feel is a little bit of cheer, a little bit of joy." Once inside the cathedral-like main hall, commuters hurry below to catch the subway, sometimes stopping to glance at playful imagery.
Play as a method for introducing light is evident at the main portion of the Angle Lake Transit Station and Plaza (p.132) in Seattle. The parking structure is cloaked in a rhythmic anodized aluminum screen, which traces the outline of the curvaceous concrete slabs. The project was inspired by the choreography methods of William Forsythe's "Dance Geometry". The sunlight piercing through this exterior dress creates streams of rays that radiate across the floors. Between the parking structure and the trains, the path is defined by a curved plaza canopy, which binds the asymmetry of the whole project to the light-rail station. Every effort is made to make what is normally a strictly utilitarian building attractive. "Parking garages are likely never to reach the noble status of a City Hall, Central Library or other important civic structures," admits lead designer, Lawrence Scarpa. "Nonetheless, they should enhance the city, the street life and the pedestrian experience, not degrade them.

那不勒斯阿夫拉戈拉高铁站，意大利
Napoli Afragola High Speed Train Station, Italy

说："停车场很可能永远达不到市政厅、中央图书馆或其他重要城市建筑的崇高地位。然而，它们应该对城市、街道生活、行人体验有着积极的提升作用，而不是降低作用。停车场的立面应该饰有建筑元素，显示其属于社区和街道财产的一部分。"

对大型建筑来说，一般认为光线通常只能到达有限的区域。然而即使像那不勒斯阿夫拉戈拉高铁站（88页）这样的大型项目，扎哈·哈迪德建筑师事务所的建筑师们仍然有办法把光线引入建筑的内部。车站蜿蜒地横跨在铁路上方，屹立在周围景观之上，看上去十分威风。但是，建筑师把它的形状进行了扭曲，减少了占地面积。其素净洁白的表面让它将自然光线温柔地吸收进来。顺着它的纵向，屋顶被拉开设计成为天窗，来给室内空间采光。整个建筑地面先是翘曲到滑入楼板的柱子里，然后流过建筑的其他部分，整个线条的起承转合如歌剧一样流畅，只有少数尖端的部分处于黑暗中。那不勒斯阿夫拉戈拉高铁站是意大利南部高速铁路网络的一个战略节点，为众多旅客和车站工作人员提供服务。这是一座繁忙的建筑，不同的楼层上发生着不同的事情，每个楼层都有不规则的形状，但是却非常明亮。项目总监Filippo Innocenti说："如果铁路全线投入运营，这个火车站每天的客流量将会达到32 700人，年客流量约1200万人。该项目还对八条铁路上面的公共人行道进行了扩建，把这条人行天桥变成了主要的客运大厅，其中设置了各种服务设施，为到达、出发和换乘的旅客提供服务，旅客可以直接从这里前往下面的所有站台。"

香港西九龙站（104页）作为通往香港的"门户"，连接着车站与周围的城市环境，地位十分重要。该项目紧靠未来的西九龙文

Their facades should be adorned with architectural elements that make them a contributing asset to the neighborhood and street."

Typically, with a large project such as Napoli Afragola High Speed Train Station (p.88), it is expected that light will be limited to certain spaces. However, even on this scale, Zaha Hadid Architects manages to draw light deep into the structure. The station snakes over the railway, straddling both sides. As a building on the landscape it is imposing, yet curls in a way that reduces its footprint, and with stark white surfaces, it gently absorbs natural light. Along its length, the roof is pulled back for a skylight to enlighten interior spaces. The fluidity with how the grade warps into columns, which slip into slabs, and then flows through the rest of the building is operatic, and leaves few sharp edges for darkness. Napoli Afragola High Speed Train Station is a strategic node in Southern Italy's high-speed network, serving a multitude of passengers and station workers. This is a busy building where a variety of activities occur on multiple levels, each bound by irregular forms, yet bright. "Once all lines are operational, 32,700 passengers are expected to use the station every day (approximately 12 million passengers each year)," project director Filippo Innocenti describes. "The design enlarges the public walkway over the eight railway tracks to such a degree that this walkway becomes the station's main passenger concourse – a bridge housing all the services and facilities for departing, arriving and connecting passengers, with direct access to all platforms below."

Acting as the "gateway" to Hong Kong, Hong Kong West Kowloon Station (p.104) was considered vital to connect the station with the surrounding urban context. The site's prominence immediately adjacent to the future West Kowloon Cultural District and next to Victoria Harbour required a design that was completely motivated

普林斯顿车站大厅和商店，美国
Princeton Transit Hall and Market, USA

化区，毗邻维多利亚港，其设计需要充分响应民众的要求。主厅的内部几乎像一片森林，倾斜的钢柱有力地支撑着屋顶和幕墙，4000块玻璃板组成的幕墙将自然光引入建筑内部。项目设计总监安德鲁·布隆伯格说："最独特的挑战是如何让地下车站看起来不像是在地下，我们花了很大的精力研究如何让自然光可持续地从富有生气的屋顶涌入地下20m的建筑层。"

多式联运枢纽的大部分结构位于地面之上，因此采光相对容易，相比之下，地铁站的照明工作就要难上很多，就好比冰山，相对于下面那部分的问题，上面看得见的问题就相形见绌了。赫尔辛基阿尔托大学地铁站（16页）的顶部表面上几乎没有被折叠。耐候钢板制成的雨篷和饰有三角形图案的天花板形成了一条连续的"丝带"，在地上街道和地下站台之间蜿蜒。其中一个立面上的斜三角形洞口将自然光引向下方。ALA建筑师事务所的塞缪尔·伍尔斯顿解释道："在西铁项目中有一点十分重要，那就是要让隧道影射当地的样貌特征而非用隧道改变城市的样貌特征。这意味着新线路沿线的每个车站都有其独有的特点。"在这个案例中，光线是上面外部城市和下面内部平台的连接点。

普林斯顿车站大厅和商店（38页）坚固的柱子支撑着一块不规则的黑色钢板，钢板向下倾斜，形成广场上方的雨篷。钢屋顶的较低部分陡峭地悬在自然光线充足的主门厅上。乍一看，与街道相连的入口非常宏伟，柱子也很壮观。然而，在建筑和广场的周围，材料和颜色选择以及细高的窗户都反映了人们对自然光的追求。建筑师里克·乔伊描述说："我们给历史悠久的普林斯顿大学

by civic demand. The interior of the main hall is almost like a forest, with leaning steel columns robustly supporting the rooftop and curtain walls laid with 4,000 glass panels to bring in natural daylight into the building. "The most unique challenge was how to make a below-ground station not seem to feel below ground," says Andrew Bromberg at Aedas, project design director. "Great effort was spent studying how to sustainably allow for a flooding of natural light, from the animated roof above, into these lower levels, located 20 meters below ground."

Whereas multimodal hubs benefit from much of their structures being above ground, making the path of light easier, subway stations are more difficult to illuminate. Similar to the composition of an iceberg, what is not seen below dwarfs what is seen above. The top of the Aalto University Metro Station (p.16) in Helsinki barely folds above the surface. The weathered steel canopy and ceiling with triangular patterns form a continuous ribbon that curls between the street level and metro platform below. On one elevation, a slanted triangular opening directs natural light downwards. "In the West metro project, we have seen it as important to push the identity of the borough into the tunnel and not to bring up tunnel identity into the city fabric," Samuli Woolston, an architect of ALA Architects, describes. "This means that each station along the new line has its own unique character." Light is here used as a connector between the exterior city above, and the interior platform below. The robust pillars of Princeton Transit Hall and Market (p.38) support a jagged sheet of blackened steel, which slices downward to form a canopy over a plaza. The low point of the steel roof hangs precipitously over the main hallway which is bathed in natural light. At first glance, approaching from the street, the entrance appears grandiose, and its pillars monumental. However, in the immediate presence of the building and the plaza, the

拉赫蒂枢纽站，芬兰
Lahti Travel Center, Finland

西港2号码头大楼，芬兰
West Terminal 2, Finland

设计了一个新入口，车站大厅的立面位于布莱尔步行街的等高线上。我们还使用四种建筑材料打造了一个简约的外观，突显了宽敞的空间和共享的公共场所。"

对公共汽车终点站来说，灯光能起到防止轻微犯罪、袭击和其他犯罪行为的作用，尤其是在黄昏之后。JKMM建筑师事务所在芬兰拉赫蒂新枢纽站的设计中优雅地运用了这一策略（50页）。公交站顶棚采用穿孔铜板反射光线，即使在晚上也能保持空间明亮，给通勤的乘客一种安全感。铜色与周围砖石建筑的色调交织在一起，即使在晚上发出夺目的光也不会影响这种整体感。该项目的主创建筑师Samuli Miettinen说："拉赫蒂枢纽站的外形如极简抽象派的雕塑，承载着历史，具有建筑上和文化上的双重价值。它创造性地提出一个易感、优质的设计理念，统一了功能复杂的城市环境。"

正如PES建筑师事务所在其2018年的项目回顾中所述，码头建筑有时缺乏适当的照明，被视为"非场所"。他们设计的赫尔辛基西港2号码头大楼（214页）和所有21世纪的码头大楼一样庞大。它的外部形式延伸到码头的道路上，天花板以俯冲的造型覆盖建筑内部。这座建筑好像是特意面对前往此处的旅客，并成为陆地和海上的灯塔。这个空间肯定是供大批旅客使用的，因此没有设计多少可以隐藏的阴暗死角。考虑到一些游轮船体较高，建筑师把码头大楼的位置抬高，以加快游客上下船的速度。总设计师Tuomas Silvennoinen表示："每年大约有400万到500万的旅客来往于西港2号码头大楼，因此设计的出发点是打造一条流畅和高

choice of materials, colors, and tall slender windows reveal that this project appeals to the human desire for natural light. "We designed a new gateway to the historical Princeton campus by placing the leading face of the Transit Hall at the contour line of Blair Walk," architect Rick Joy describes, "Four materials were used for simplicity, emphasizing a generosity of space and collective public areas."

For bus terminals, light is used to prevent petty crimes, assaults, and other criminal behaviour, particularly after dusk. JKMM Architects uses this strategy with elegance for Lahti Travel Center (p.50). The reflective nature of the perforated copper of the bus canopy helps to keep the stations bright even at night, giving commuters a feeling of security. The color of the copper blends with the surrounding stone and brick-clad buildings, even as it brings a blinding luster to the area. "The minimalistic sculpture-like form of Lahti Travel Center embraces the history and value of the culturally and architecturally significant milieu," says Samuli Miettinen, the principal architect of the project. "It creates an easily perceivable and high-quality design concept, which unifies a functionally complex city environment."

Sea terminal buildings sometimes lack proper lighting and are considered "non-places", as PES-Architects states in the project reviews of 2018. West Terminal 2 (p.214) in Helsinki is as voluminous as any 21st century airport terminal. Its exterior form stretches across the roadway of the pier, while its ceiling covers the interior in a swooping fashion. It is as if the building is deliberately turning to face its oncoming users, to be a beacon of light by land and by sea. This is definitely a space for a mass of people, with few dark corners to hide. The terminal shows deference to the ships with an elevated posture, which speeds up the embarkation and disembarkation processes. "Around 4-5 million ferry passengers a year travel through West Terminal 2, and the

里斯本游轮码头，葡萄牙
Lisbon Cruise Terminal, Portugal

波尔图游轮码头，葡萄牙
Porto Cruise Terminal, Portugal

效的客流通道。我们想为到达赫尔辛基的旅客创造一个优雅而独特的大门，相信这一点已经通过该建筑的形式得以实现，白色的立面在阳光下闪闪发光，两边弯曲着朝向地面，让人想起一些被冲上岸的海洋生物。"

里斯本游轮码头（178页）的动态极简主义风格给人的印象是它真的在努力地保持低调。其造型谦逊到几乎像是在表达歉意，仿佛它从地面升起，只是为了与舷梯处于同一高度。人们享受着这座俯卧在塔古斯河口的码头带来的好处，码头同时也成为另一些站在阿尔法玛山坡上的人们眼里的风景。室内通风良好，但通过大型的窗户和倾斜的屋顶板使阳光充分入射；而室外则更加有趣。室内宽敞的空间加快了旅客的移动速度，而室外镶软木混凝土结构又激发了他们的好奇心。Carrilho da Graça建筑事务所的建筑师热情地说道："一条小路或者说走廊围绕着建筑，让人们一边慢慢探索周围的环境，一边经过不同的建筑立面。这条小路一直通到屋顶为止，屋顶呈现出一种舞台感，如同一座广场，将河流与城市毫无障碍地连接起来。"游轮码头就是这样一座好像很害羞的建筑，越是试图避开人们的注意，越是引得人们想去寻找它的存在，直到来到楼顶对其确认。

波尔图游轮码头（194页）由路易·佩德罗·席尔瓦建筑公司设计建造，码头紧挨着水面，一边为大型船只预留了空间，另一边为小型船只预留了空间。六边形的陶瓷片布满整个立面和一部分内墙，甚至在舷梯上也绕了一圈，使整座建筑从里到外都闪闪发亮，在海上十分显眼。圆形的楼层平面鼓励旅客走动，欣赏建筑美丽的光泽，而乐师则在中央大厅弹奏着马托西纽什小调迎接他

starting point for the design was to ensure smooth and efficient passenger flows," describes the chief designer, Tuomas Silvennoinen. "We wanted to create an elegant and distinctive gate to Helsinki for arriving passengers. We believe this has been achieved in the form of the building, with white facades gleaming in the sun and the sides curving towards the ground, reminiscent of some sea creature washed ashore."

The dynamic minimalism of Lisbon Cruise Terminal (p.178) gives off the impression of a building that is trying really hard not to make a scene. Its form is almost apologetic, as if it rises from the ground only to be on the same level as the network of gangways. The people who benefit from the terminal's crouched position are those with views of the Tagus estuary from the Alfama slope. In contrast to an interior that feels airy, but uses expansive windows and tilted roof slabs to optimize the entry of sunlight throughout the day, the exterior is more intriguing. The spaciousness of the interior speeds users up, while the concrete with cork exterior stimulates curiosity. "A path/promenade surrounds the building, allowing for a slow discovery of the surroundings while passing through the different facades," Carrilho da Graça Arquitectos eloquently describes. "This path culminates in the roof that assumes the features of a stage, relating with the river and the city without any obstacles, like a plaza." The terminal is such a shy building, trying to evade detection as much as possible, that it prompts users to investigate its existence, which is confirmed when they reach the top.

The Porto Cruise Terminal (p.194) by Luís Pedro Silva, Arquitecto is furled up next to the water, with space for larger ships on one side, and space for smaller vessels on the other. The hexagonal scales on the facade, some interior walls, and twisting over the gangway, glisten throughout, reflecting the building's maritime location. The circular floor plans encourage passengers to ambulate, appreciating the sparkling beauty of the building,

停车楼和屋顶，丹麦
Parking Houses + Konditaget Lüders, Denmark

们。无论在什么地方观察，码头给人的视觉体验在晚上都是最大胆的。该公司网站上有这样的描述："从远处瞭望，它洁白蜿蜒的体量细致入微地描绘出光线与大气的变化。"这个游轮码头给旅客们留下了闪亮的第一印象或是最终印象。

如果海上的码头大楼往往是阴暗的，那么停车库就因缺乏光线而饱受诟病。然而，停车库给建筑师提供了许多自由发挥的空间，施展他们应用光线的才华。JAJA建筑师事务所设计的哥本哈根停车楼和屋顶（148页）就利用了空间上的自由度，告别了依靠结构把内部气候环境包起来的传统封闭结构，让停车场的立面可以呼吸。立面中唯一的实心部分和外边的楼梯平行，弯曲的楼梯像胳膊一样环抱着大楼。多功能结构的屋顶给各个年龄段的人提供了娱乐的场地。Kathrin Susanna Gimmel作为负责该项目的主要建筑师之一，认为公园的开放与可达性与城市息息相关。她做出了解释："该项目的任务是做停车场结构立面和屋顶的设计，将其整合到城市结构中的同时也回馈给城市一些好处。所以停车场不再仅仅是一个占位空间，而是能够使自己的空间派上一些用场，为大家服务的地方。"

还有一个独具匠心的建筑案例就是荷兰的卡特韦克海滨地下停车场（164页）。它给人的第一感觉就像一些藏在沙丘中的小型帐篷。地面上的各个入口与通向地下停车场的楼梯相连，从某些角度你几乎看不见它们。设计师没有让新建的停车场给卡特韦克宝贵的海滨风光留下"伤痕"，而是给世界上其他地区的海滨圣地提供了一个发人深省的范例。Royal HaskoningDHV事务所的

while musicians greet them in the central foyer with the local tunes of Matosinhos. The sensual merriment of the terminal is most audacious at night, no matter where it is observed. According to the firm's website: "From far away, the building is read by its volumetric and wavy white(ness) with fanciful nuances regarding light and atmosphere variation." The terminal makes both a shiny first and last impression for its users.

If sea terminals tend to be dark, parking garages are notorious for their lack of light. Yet, the freedom that parking garages offer to architects is the ability to handle light more liberally. Absent the requirement to enclose, which forces other structures to protect internal climatic conditions, the facade of a parking garage is allowed to breathe. The Parking Houses + Konditaget Lüders in Copenhagen (p.148) by JAJA Architects takes advantage of this freedom, by using perforated surfaces on the elevation. The only part of the facade that is solid is parallel to the exterior stair, which twists and hugs the building like an arm. The rooftop space is a multigenerational playground for a multifunctional structure. According to Kathrin Susanna Gimmel, one of the main architects who worked on the project, Park "n" Play's openness and accessibility is what relates it to the city. "The task was to make a facade and rooftop design that integrates the parking house in the city structure, but at the same time also gives something back to the city," she explains. "So the parking house doesn't just take space, but it gives space back to users, accessible for everyone."

Another example of Dutch ingenuity is Underground Parking in Katwijk aan Zee (p.164). At first impression, it resembles a collection of small pavilions hidden within sand dunes. The entrances to the stairs which lead to the subterranean parking garage peak above the surface, and from certain angles are barely seen. Rather than devote precious beachfront scenery to the blight of a parking lot, the Katwijk aan Zee is a thoughtful example

卡特韦克海滨地下停车场，荷兰
Underground Parking in Katwijk aan Zee, The Netherlands

Richard van den Brule解释道:"位于卡特韦克海滨的地下停车场是库斯特沃克·卡特韦克项目的一部分，该项目旨在展示荷兰在未来海岸线加固工程上的创新做法。在保护海岸线的同时，我们还朝大海方向拓宽了海滩面积，方便市民和游客更充分地享受海滨生活。"游客们踩着小径和木板道在地面上闲逛，有些人对脚下的基础设施浑然不觉，直到他们看到一条向下的坡道才发现它们的存在。因此，他们可以享受海滩上的阳光，几乎不会受到结构上的干扰。

接着我们将目光投向市政项目。长期以来，它们一直被视为"不需要见光的地方"，可对纽约的克罗顿滤水厂（228页）来说，建筑的功能并没有被隐藏在一个毫无特点的混凝土"盒子"里，而被隐藏在一个椭圆形的小型高尔夫球场下面，球场一侧围着护网来接球。就像护网接住飞过来的球一样，球场上的水池、沟渠、湿地也可以接住从天而降的雨水，并将其过滤。滤水厂能为纽约市民提供"超过1万亿吨的纯净水"。任务如此繁重的工程经过精致的表面处理，反而显得更加轻盈和开放。

一个又一个的项目向我们展示了运用光线可以使城市基础设施更加人性化，改变以往人们对基础设施的印象，让他们觉得基础设施不再是城市的"伤痕"，而是优秀的建筑作品。正如我们所见，各种类型的基础设施都真实地反映了这些转变，其中光这个元素以独到的方式被运用，产生不同的独特效果。总的来说，我们的城市基础设施最终会变得更具魅力，我们的城市也会因为扩建变得更加迷人。

that other popular beaches in the world should study. "The Underground Parking in Katwijk aan Zee is part of 'Kustwerk Katwijk', a project which illustrates how the Netherlands has to be innovative in its approach to reinforcing its coastline for the future," explains Richard van den Brule of Royal HaskoningDHV. "In addition to protecting the coastline, the beach was expanded seawards so that the municipality and visitors could continue to enjoy the seaside." Visitors saunter about the grounds, through trails and boardwalks, some oblivious to the infrastructure beneath their feet, until they come upon a sunken roadway. As a result, they can enjoy the sunlight on the beach with little structural interruption.

Then there are the municipal projects, which for too long were seen as places that did not need light. For the Croton Water Filtration Plant (p.228) in New York, the functions are not hidden within a faceless concrete box. Instead, the functions are hidden below a tiny, oblong golf course, with nets draped on one side to catch drives. As the nets capture balls, the basins, moats, and bioswales capture rainwater which gets filtered. The plant is capable of providing New Yorkers with "over 1 trillion litres of cleaned water". This is a heavy site, but through delicate surface treatments it appears lighter and more open.

Project by project, light is used to humanize urban infrastructure, changing the perception of them from urban blights to celebrated works of architecture. As we have seen, this transformation is true across infrastructure types, where light is used in disparate ways to achieve specific results. The overarching result is that our urban infrastructure has finally become more appealing, and by extension, the quality of our cities has too.

盎格鲁湖中转站和广场是基于多功能的理念创建出来的基础设施，集生活、工作及娱乐等功能于一体，为人们提供了充足的空间，它包含了4000m²的广场、社区活动空间、轻轨落客区、零售商铺、配有集成车锁和架子的专用自行车存放处以及电动车充电站。项目中还包括一块3250m²的空地，尚待开发成为交通用地，此外一座能够容纳1150辆汽车的停车场，在未来会根据使用需求进行下一步的改造。盎格鲁湖中转站每天要服务超过2500名乘客，因此成为城市交通设施中的一个重要枢纽。

中转站的设计灵感来源于威廉·弗西斯的即兴作品《舞蹈几何》，在该作品中，舞者们根据空间中折叠弯曲的线条来扭动身姿。建筑师认为简单的直线能够形成无限的动作和位置，需要的过渡也是最少的。这一想法能够促使人们不过多地去考虑最终结果，而是专注于发现过程中新的动作和变换形式。

这个占地面积为3.7ha的多功能建筑群出自国际建筑设计竞赛的优胜建筑师之手。建筑的主体是七层的后张拉混凝土结构，采用现场浇筑法浇筑。外立面由7500块定制的蓝色阳极氧化铝立面板组成。遵从平面几何学原则，建筑波浪起伏的立面由两条曲线和连接二者的一系列直线构成。每一个定制的铝立面构件都采用结构简单、尺寸标准的设计，形成有效的结构形状和材料形式，也最大限度地提高了生产、制造、安装过程中的成本效益。整个立面的装配工作仅仅用了三周的时间，且没有借助起重机等专门设备。

建筑师充分利用了场地的倾斜地势，在这座多功能的建筑中设计了三个入口，这三个入口都朝向同一条街道，但高度不同。建筑有五层位于地上，另外两层有一部分匿于地下。建筑首层包含232m²的零售空间，西侧则是3251.6m²尚待开发的交通用地。位于四层的公共广场直接连通轻轨入口处、停车场和公共街道，使各部分在物理空间和视觉

盎格鲁湖中转站和广场
Angle Lake Transit Station and Plaza

Brooks + Scarpa

上都彼此连接。广场内包括一个落客区、辅助客运负载区，以及由车库通往车站的带顶走廊，走廊中展示了与当地特色有关的艺术作品。具有装饰性的座椅墙、通道、铺地材料、当地花草以及集水装置共同为在车站中行走和等待的人们提供了安静的社交空间。广场还为社区活动提供了场所，人们可以在此进行如节日庆典、农产品集市、艺术展览等各类室外集会。盎格鲁湖中转站和广场各主要部分的设计和位置在实现空间有效利用的同时，也最大限度地实现了建筑的功能性、可持续性和美观性。

With ample space for people to live, work, and play, the new Angle Lake Transit Station and Plaza is an envisioned mixed-use facility consisting of a 4,000m² connecting plaza and community event spaces, a drop-off area for light rail users, retail space, dedicated bike storage with integrated lockers and racks, and charging stations for electric vehicles. It also includes a 3250-square-meter parcel for future transit-oriented development and a parking structure for 1,150 cars, designed to accommodate a conversion upon new future uses. Serving over 2,500 passengers daily, Angle Lake Transit Station is an important transport hub in the Sound Transit portfolio of transit facilities.

Inspired by William Forsythe's improvisational piece "Dance Geometry" where dancers connect their bodies by matching lines in space that could be bent, tossed or otherwise

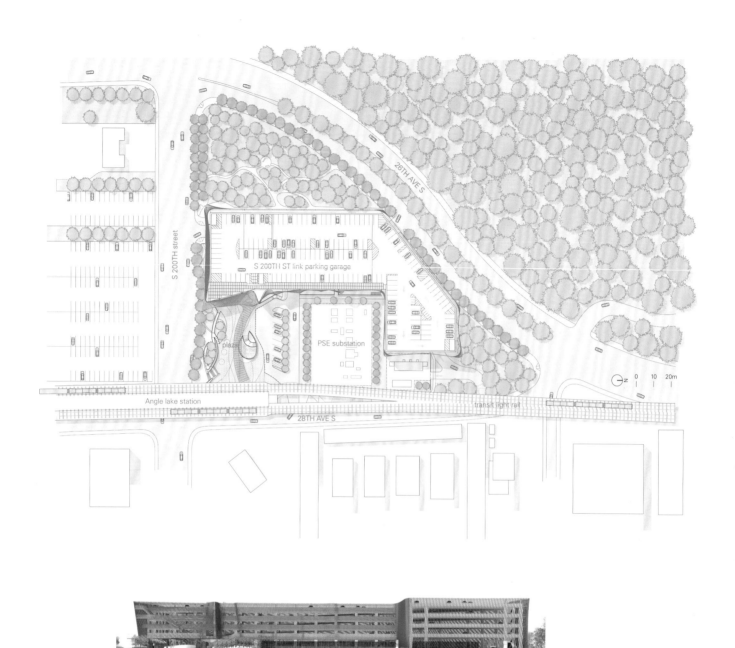

东立面 east elevation

北立面 north elevation

南立面 south elevation

西立面 west elevation

courtesy of the architect

1. 零售区域	27. 公共室	15. VMS(variable measure sign)
2. 自行车存放处	28. 储藏室	16. signal at garage entry
3. 景观	29. 安全屏幕	17. pay phone
4. 连接不同楼层的楼梯	30. 公交站	18. women's restroom
5. PSE	31. 公交站遮阳棚	19. men's restroom
6. 临时停靠接送区	32. TPSS	20. janitor
7. 辅助客运系统落客区	33. 大屏幕	21. future payment stations
8. 雨篷	34. 高承载车辆停放区	22. electrical room
9. 艺术项目	35. 低排放车辆停放区	23. UPS
10. 中转站	36. ZIP	24. machine room
11. 坡道	37. 海湾捷运	25. mechanical room
12. 水景		26. fire sprinkler
13. 天然花草	1. retail	27. comm
14. 生态滞留沼泽池	2. bike storage	28. storage
15. 可变测量标志	3. landscape	29. security screen
16. 车库入口信号	4. cross stairs between levels	30. bus stop
17. 付费电话	5. PSE	31. bus shelter
18. 女士卫生间	6. kiss and ride	32. TPSS
19. 男士卫生间	7. para-transit drop off	33. screen
20. 门卫	8. canopy above	34. HOV(high occupancy vehicle)
21. 未来付费车站	9. art projects	35. LEFE(low emission fuel efficient)
22. 电气室	10. transit station	36. ZIP
23. UPS	11. ramp	37. sound transit only
24. 机房	12. water feature	
25. 机械室	13. natural planting	
26. 防火喷淋装置	14. bio-retention swale	

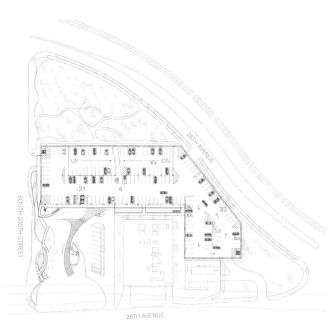

四层 third floor

一层 ground floor

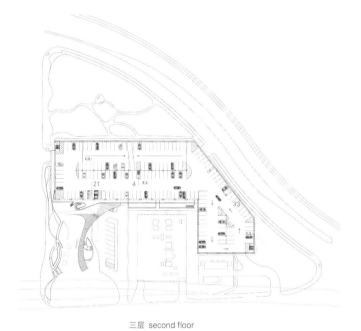

三层 second floor

地下一层 first floor below ground

二层 first floor

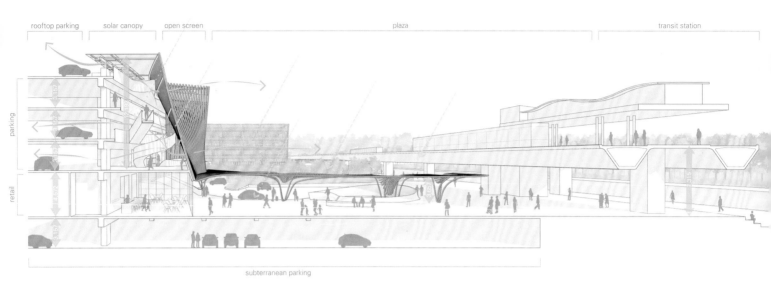

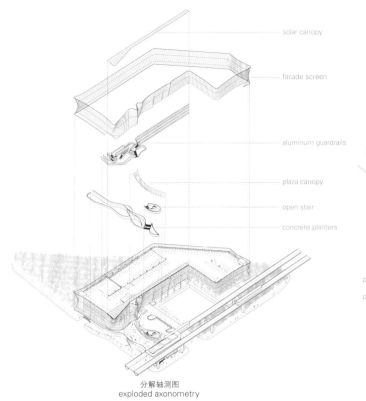

分解轴测图
exploded axonometry

- solar canopy
- facade screen
- aluminum guardrails
- plaza canopy
- open stair
- concrete planters

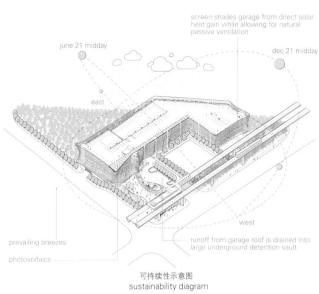

可持续性示意图
sustainability diagram

screen shades garage from direct solar heat gain while allowing for natural passive ventilation

june 21 midday
dec 21 midday
east
west
prevailing breezes
photovoltaics
runoff from garage roof is drained into large underground detention vault

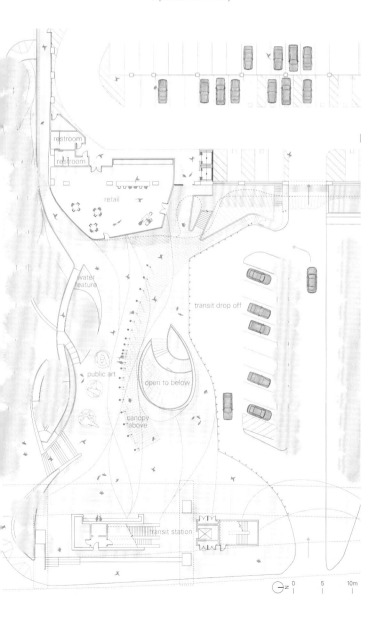

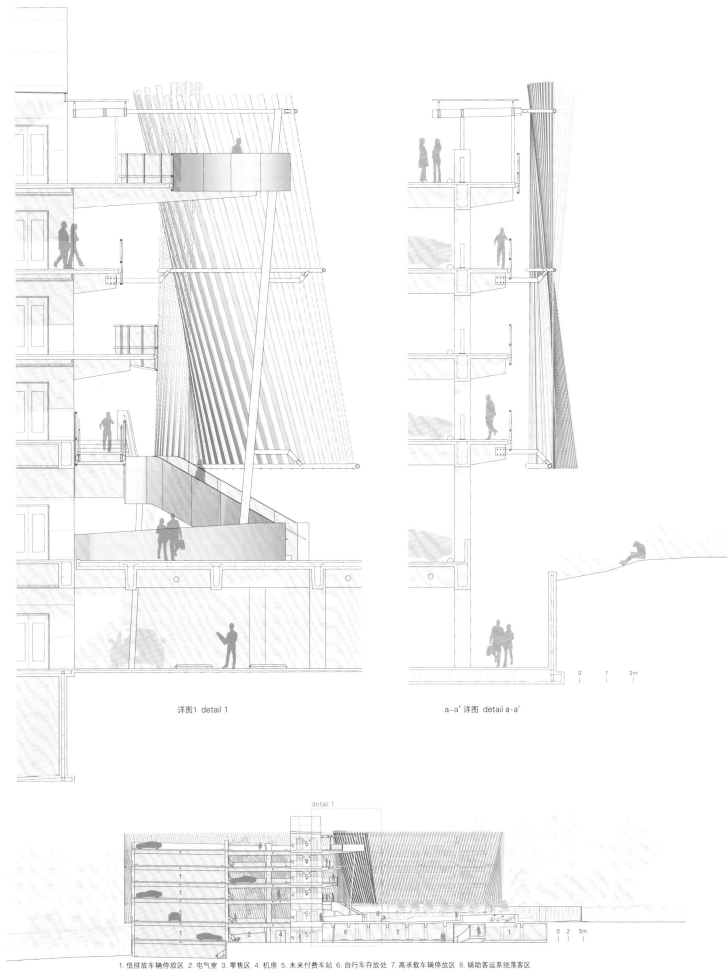

详图1 detail 1

a-a'详图 detail a-a'

1. 低排放车辆停放区 2. 电气室 3. 零售区 4. 机房 5. 未来付费车站 6. 自行车存放处 7. 高承载车辆停放区 8. 辅助客运系统落客区
1. LEFE (low emission fuel efficient) 2. electrical room 3. retail 4. machine room
5. future payment stations 6. bike storage 7. HOV (high occupancy vehicle) 8. para-transit drop off

A-A'剖面图 section A-A'

项目名称：Angle Lake Transit Station and Plaza / 地点：19955 28th Ave South, Seatac, WA 98188, USA / 事务所：Brooks + Scarpa / 当地工程师与建筑师：Berger ABAM / 项目团队（Brooks + Scarpa）：Lawrence Scarpa - lead designer; Angela Brooks - project executive; Mario Cipresso - project architect; Emily Hodgdon, Mark Buckland, Jeff Huber, Chinh Nguyen, Cesar Delgado, Fui Srivikorn, Christina Wilkinson, Royce Scortino, Sheisa Roghini, Soha Momeni, Ryan Bostic - project design team / 项目团队（BergerABAM）：Bob Griebenow - project executive; Lars Holte - project director + engineer Landscape: Brooks + Scarpa; David Sacamano - BergerABAM / 工程师：BergerABAM - structural + civil engineering; Stantec - electrical and lighting; Sazan Group, Inc. - mechanical; Luminescense - lighting design; Stantec - security; Shannon & Wilson - geotechnical / 承包商：Harbor Pacific/立面工程师：Brooks + Scarpa, Lars Holte, P.E., Walter P. Moore / 立面生产商：APEL Extrusions and Intermountain Industrial Fab / 客户：Sound Transit / 造价：$36.1 million / 竣工时间：2017 / 摄影师：©Ben Benschneider (courtesy of the architect) (except as noted)

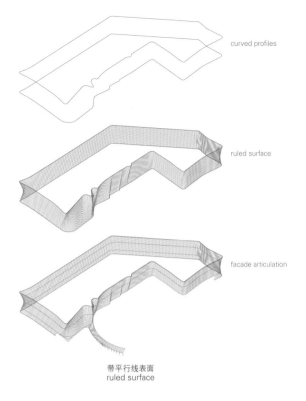

curved profiles

ruled surface

facade articulation

带平行线表面
ruled surface

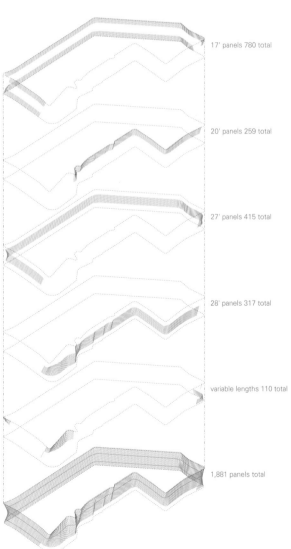

17' panels 780 total

20' panels 259 total

27' panels 415 total

28' panels 317 total

variable lengths 110 total

1,881 panels total

拼接表面
splicing study

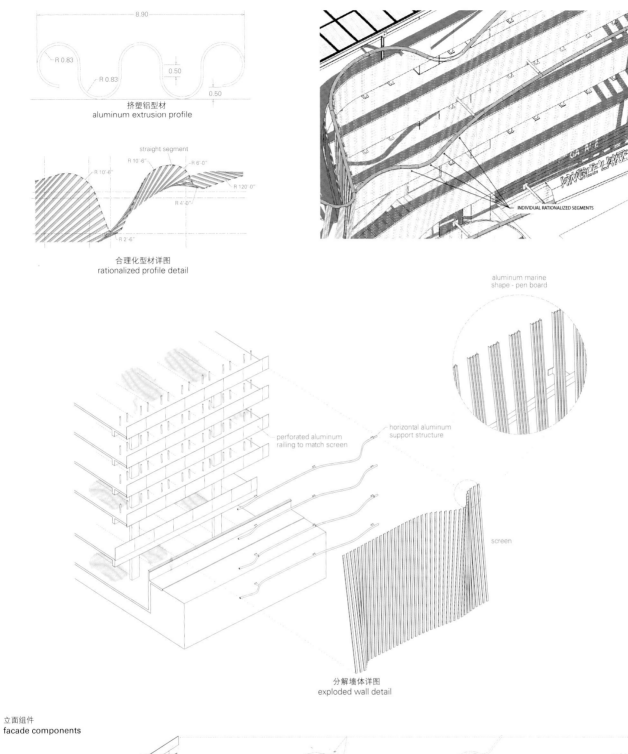

挤塑铝型材
aluminum extrusion profile

合理化型材详图
rationalized profile detail

分解墙体详图
exploded wall detail

立面组件
facade components

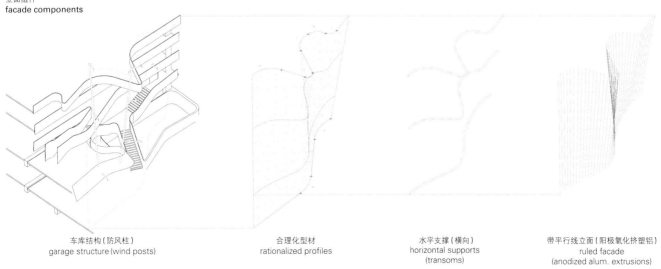

车库结构（防风柱）
garage structure (wind posts)

合理化型材
rationalized profiles

水平支撑（横向）
horizontal supports
(transoms)

带平行线立面（阳极氧化挤塑铝）
ruled facade
(anodized alum. extrusions)

distorted, architects began to think of the possibilities how simple straight lines can be composed to produce an infinite number of movements and positions with a minimum need for transition. This idea lessened the obligation to think about the end result thus enabled to focus more on discovering new ways of movement and transformations in the process.

The 3.7ha multi-modal transit complex was the result of an international design/building competition. It features a seven-storey, cast-in-place and post-tensioned concrete structure with an exterior facade that uses over 7,500 custom formed blue anodized aluminum facade panels. Using ruled surface geometry, the undulating facade is formed by connecting two curves with a series of straight lines. Each of the custom aluminum facade elements was designed and segmented into a simple, standardized size in the most efficient structural shape and material form while maximized the cost-efficiency in production, fabrication, and installation. The entire facade was installed in less than three weeks without the use of cranes or any special equipment.

Taking advantage of the sloping site topography, this multi-functional structure is accessible from three different locations on the street at various levels. Within five levels above ground and two levels partially below ground, 232 square meters of retail space locates at ground level and a 3251.6-square-meter site for future transit-oriented development sits on the west. The public plaza, on the third level, connects directly to the entrance of the light rail, parking structure, and public streets, forming a physical and visual connection between the project elements. It includes a passenger drop-off area, para-transit loading areas, and a covered walkway from the garage to the station, displaying several regionally inspired artworks. Ornately designed seat walls, pathways, paving, native planting, and storm-water catchment features create quiet places for social interaction while waiting for a transit connection and help transit users to engage as they move through the space. The plaza is designed to also accommodate community events, such as festivals, farmers' markets, art exhibits, and other outdoor public gatherings. The design and location of major elements in Angle Lake Transit Station and Plaza together seek to maximize function, sustainability, and aesthetics while providing an efficient use of space.

停车楼和屋顶
Parking Houses + Konditaget Lüders
JAJA Architects

应该说，停车楼是哥本哈本市不可或缺的一部分。但是我们该如何使停车楼摆脱只有停车这一项职能的传统？如何使停车场在实现多功能化的同时成为富有魅力的公共空间？如何创建一个既符合哥本哈根诺德哈文开发新区的规模，又尊重城市历史文化和未来愿景的大型停车楼？

Århusgadekvarteret这个地区历史悠久，拥有特色鲜明的红砖海港建筑，被称为"红色街区"，是诺德哈文大开发计划的第一环，会在其历史面貌的基础上，在现有的特征上被赋予新的诠释。竞赛项目的地点就位于红色街区，是由客户提供的一个传统的停车场。竞赛的任务是创建一个引人入胜的绿化立面，并鼓励人们去使用屋顶空间。

传统意义上的停车场通常被隐藏在建筑内部，建筑师没有采用这种传统的结构，他们想要在合理范围内突显网格结构那种粗犷的美感，因此要打破和重新组合大面积的整体立面。由于整座建筑与周围的城市空间很接近，建筑师必须想办法避免其庞大的体型从近处来看呈现出那种突兀感。因此他们将其设计成一个红色的混凝土结构，作为该地区特色红砖的自然延续。灰色框架被涂红，显得温暖而亲切。由"植物架子"组成的系统使植物分布在整个建筑立面上，该系统的布置与立面网格一致，从而引入了一种新的规模。建筑师以结构与自然环境的互动为基础来表达立面设计，而结构与自然环境的互动也是一种令人兴奋的相互依存关系。立面上由植物种植箱组成的网格被两部大型公共楼梯截断。主动型停车楼设计的基本原则是为当地居民和游客提供一个无障碍且具有娱乐性的屋顶。因此，在创建生动的屋顶时，可见性和无障碍性至关重要。楼梯参考了蓬皮杜文化中心的楼梯设计，使得人们沿着立面移动本身就是一种体验。建筑师们和RAMA工作室合作，在楼梯上设计了带状的围栏，如同现代神话一般讲述了这个港口地区的工业历史，并展望了其作为哥本哈根开发新区的未来愿景。

上楼时楼梯的围栏只是起到扶手的作用，人们顺着扶手来到屋顶，就可以看到扶手变成了秋千、球护圈、丛林体育馆等各种趣味游乐设施。红色的扶手仿佛拉着游客们的手，邀请他们从街道层来到屋顶，享受这里的愉快氛围，欣赏哥本哈根港的美景。在屋顶，扶手成了一座雕塑，提供了令人兴奋的体验，还形成了休息空间和游戏空间。雕塑穿过屋顶后没有中断，沿着第二部楼梯又返回街道层。

像这样，这些围栏成为贯穿整座建筑的红线，使建筑立面、楼梯和作为活动空间的屋顶连成一个整体。哥本哈根的新停车楼将成为一个社交场所，也将成为展现当地环境活力的一个平台，对于这个城市的居民、运动者和游客们来说是很好的礼物。

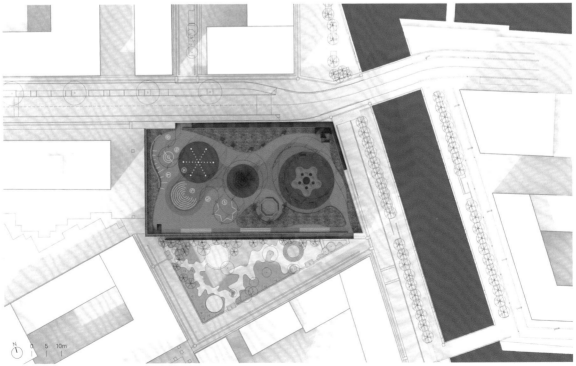

Parking houses should be an integral part of the city. But how can we challenge the monofunctional use of the conventional parking house? How do we create a functional parking structure, which is also an attractive public space? And how do we create a large parking house that respects the scale, history and future urban culture of the new development area Nordhavn, Copenhagen?

Århusgadekvarteret, known as the Red Neighborhood because of the historical character of red brick harbour buildings, will undergo the first phase of a major development plan for Nordhavn to build upon the historical trait and merge existing characteristics into new interpretations. The starting point for the competition project was a conventional parking house structure in such area. The task was to create an attractive green facade and a concept that would encourage people to use the rooftop.

Instead of concealing the parking structure, which is a normal practice, the proposed concept enhances the beauty of a rational crudeness of the structural grid, while breaking up the scale of the massive facade. Because of its proximity to the surrounding urban spaces, the large volume had to avoid being predominately seen from close-up. As a natural continuation of the area's red brick identity, the grey frame was colored red, thus radiates warmth and intimacy. A system of "plant shelving" distributes greenery across the entire facade in a rhythm relating to the grid and introduces a new

基本结构 basic structure

红色结构 red structure

绿色立面 green facade

活跃的屋顶 active roof

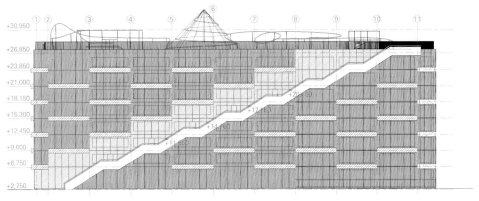
南立面 south elevation

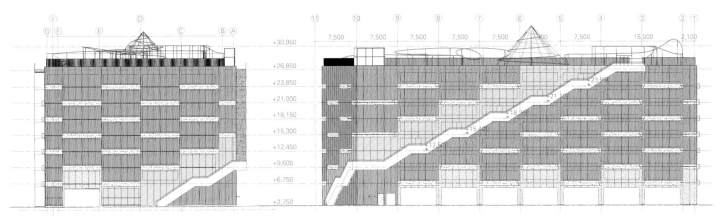
东立面 east elevation　　　　　　　　　　北立面 north elevation

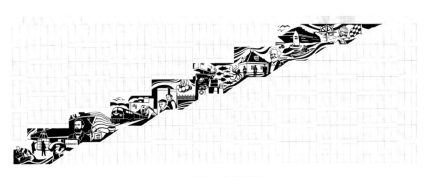

南立面——带状示意图
south facade _ frieze illustration

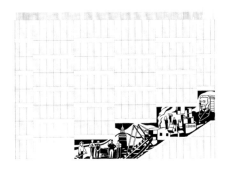

东立面——带状示意图
east facade _ frieze illustration

北立面——带状示意图
north facade _ frieze illustration

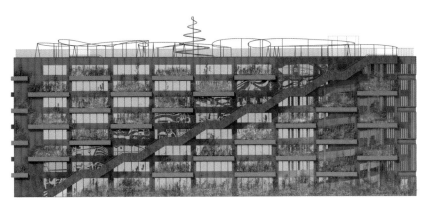

南立面 south elevation

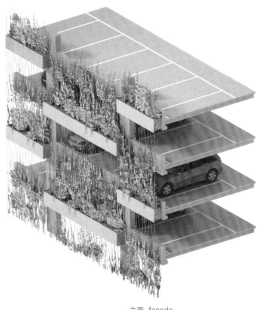

立面 facade

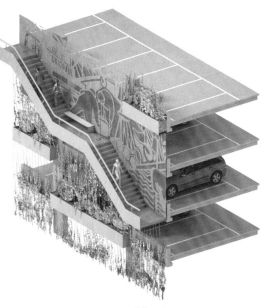

楼梯 staircase

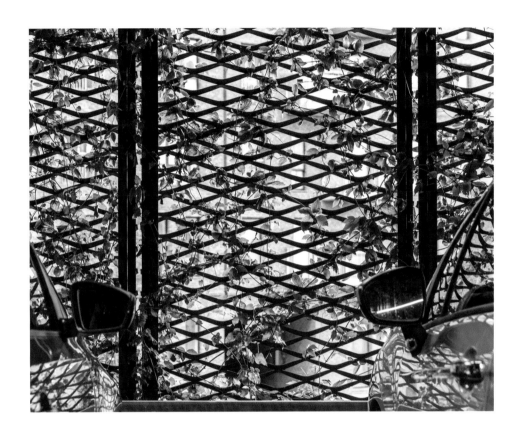

屋顶 roof

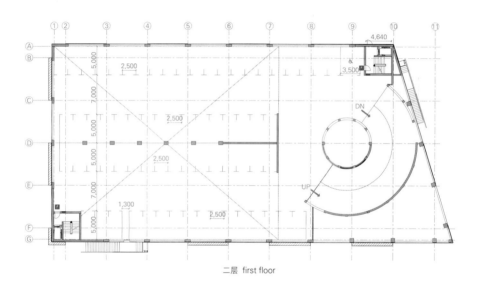

二层 first floor

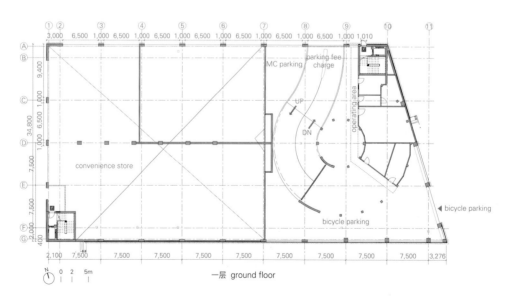

一层 ground floor

项目名称：Parking Houses + Konditaget Lüders / 地点：Copenhagen, Denmark / 事务所：JAJA Architects
合作设计团队：Søren Jensen Ingeniør, RAMA Studio, LOA, DGI / 合作承包商团队：5e byg, Aarstiderne Arkitekter, Ingeniør'ne / 客户：CPH City & Port Development / 功能：green facade and activity landscape on roof / 面积：2,400m² (roof), 4,800m² (facade), parking for 485 cars + 10 MC / 服务：main consultant on architecture and landscape / 设计开始时间：2014 / 竣工时间：2016
摄影师：©Rasmus Hjortshøj–COAST (courtesy of the architect)

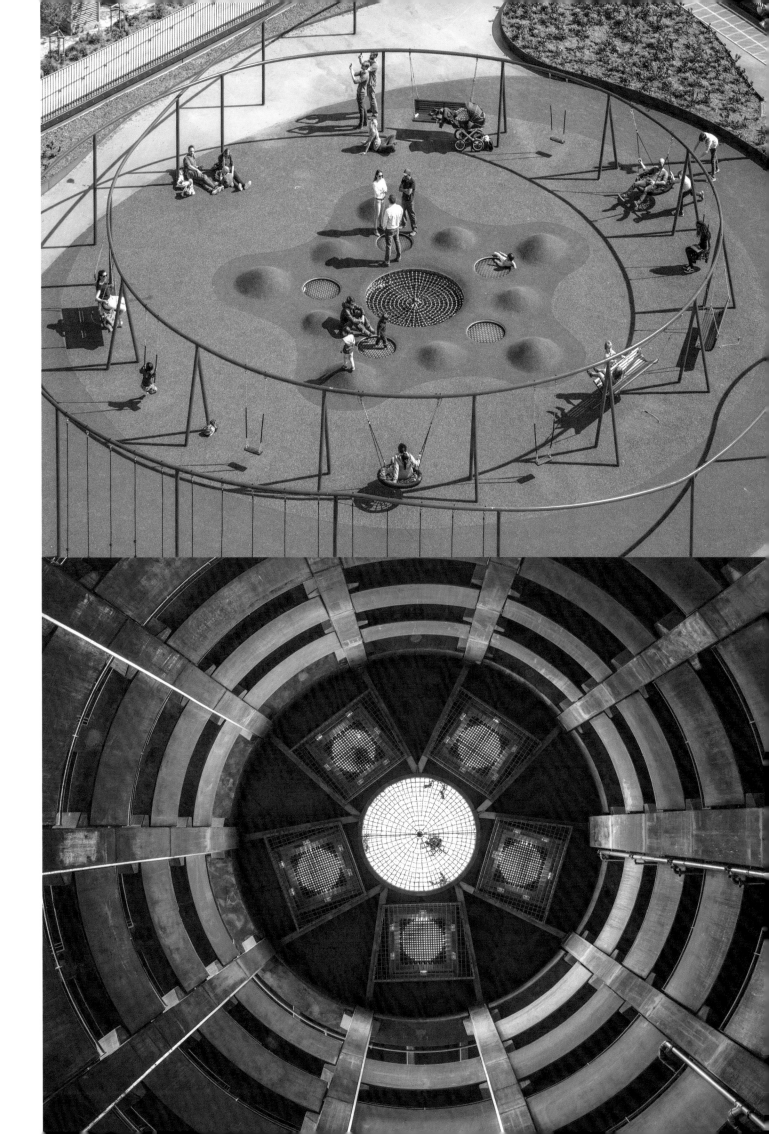

scale. The expression of the facades is based on an interaction between structure and nature – an exciting interdependence between the two. The grid of plant boxes on the facade is then penetrated by two large public stairs. The basic principle of an active parking house was to offer a recreational roof to local inhabitants and visitors. Visibility and accessibility were therefore essential when creating a living roof. The staircase has references to Center Pompidou, where the movement along the facade is an experience in itself. In collaboration with RAMA Studio, the graphical frieze was created on the stairs to become a modern tale of the harbour area's industrial history and its future as Copenhagen's new development. From being a mere railing, it transforms to become a fantastic playground on the rooftop – swings, ball cages, jungle gyms and more. From street level, the railing literally takes the visitors by hand, inviting them on a trip to the activity landscape and amazing view of the Copenhagen Harbour. Here, it becomes a sculpture and offers exciting experiences, resting spaces, play areas, and spatial diversity. The sculpture's journey across the roof continues uninterrupted, before leading back along the second staircase towards the street.

As such, the structure becomes a red thread through the project, joining the facade, the stairs, and the roof for activities as one single element. Copenhagen's new parking house will be a social meeting ground and an active part of its local environment – as an urban bonus for locals, athletes and visitors alike.

卡特韦克海滨地下停车场
Underground Parking in Katwijk aan Zee

Royal HaskoningDHV + OKRA Landscape Architects

在荷兰,超过四分之一的土地位于海平面以下,因此加固海岸线、抵御水患就显得极为必要。卡特韦克海滨地区及其防洪堤都存在安全隐患,因此政府一直在推进库斯特沃克·卡特韦克项目,保护卡特韦克的海岸线,同时增设一些停车场所,缓解这一区域的停车问题。受市政府委托,Royal HaskoningDHV事务所负责地下停车场的设计工作。同时,车库的布局,旅客入口和车辆入口的布局,紧急出口的布局以及路线、方向和定位的标识设计也一并由他们全权处理。

在大海和村庄之间的堤坝上,覆盖着长达1200m、高出海平面10~11m的沙丘。沙丘不仅仅是一种隐性的防洪设施,它下面还隐匿着一个宽30m、长500m的地下车库,和经过加固的堤坝平行,从堤坝位置开始向内陆延伸。设计师将停车场结构塞进距离堤坝仅3m的沙丘中,这样可以增加海滨停车场的容量,并将两个项目合并成一个项目。设计师将663个车位的停车场精心嵌入自然沙丘环境中,在融合公共空间和自然地貌的范例中又添了新的一笔。沙丘上种植了沙茅草(用以固沙)、铺了小石径、修了木板路,还设了休息长椅,这样游客们就可以沿着沙丘漫步或前往海滩。这些沙丘形状各异,跌宕起伏,形成一个个精妙隐蔽的出入口。设计师精心塑造沙丘的形状,确保沙丘地貌特征不受破坏,同时让自然光照射到地下区域,便于人们在地下长而矮的停车场内定位。每个入口的造型都是一样的,但在位置上需要与各自的车道形成独特的连接。在停车场靠海边一侧有紧急出口,它们都是最小化设计,而且被设计成沙丘里的雕塑状结构。车库的路面由互锁式地砖拼成,为了尽可能多地采光,游客入口从上到下都是透明的。

恶劣的天气条件要求设计师使用高质量和易养护的装饰面,因此设计师们采用玻璃、(穿孔耐候)钢、瓷砖和裸露混凝土等主要材料。为了减少对环境的干扰,设计师缩短了建造停车场的时间,并降低维护成本,采用预制混凝土模块化施工,保证了墙、柱和屋顶持续的高质量。此外,许多构件为标准化设计,有着相同的细节,这使得建筑外观显得非常适度。总的来说,设计师将色彩和材料无缝地融合到卡特韦克海滨的城市肌理中,使卡特韦克海滨地下停车场因此成功地达成了项目要求加固海岸和增加停车位数量的目标,符合海滨地区景观设计的理念。

Over a quarter of the Netherlands' land mass lies beneath sea level, making it imperative to protect the country against the sea. The coastal region of Katwijk and its dike did not meet the safety standards and thus has been implementing Kustwerk Katwijk, a project which seeks to protect the coastline of Katwijk aan Zee. Additionally, the surrounding urban situation required additional parking facilities. Commissioned by the Municipality of Katwijk, Royal HaskoningDHV was responsible for not only the architectural design of the underground parking, but also the layout of the garage, the entrances for visitors and vehicles, the emergency exits, as well as the signage for routing, orientation, and identification.

Between the sea and the village, a 1,200m long sand dune tops the dike, creating a height of between 10 to 11m above mean sea level. This sand dune conceals more than just the municipality's coastal protection. A 30m wide, 500m long parking garage is located within the dune, extending parallel to – and inland from – the reinforced dike. By tucking the

1. 面向机动车道的南入口 2. 面向Andreasplein的主入口 3. Schelpenpad路 4. 逃生楼梯
5. 面向林荫大道的主入口 6. 面向Voorstraat路的主入口 7. 面向Waaigat路的北入口 8. 面向Wihelmina路的北入口 9. 面向机动车道的北入口
1. south entrance to motorized traffic 2. main entrance to Andreasplein 3. Schelpenpad 4. emergency staircase
5. south entrance to Boulevard 6. main entrance to Voorstraat 7. north entrance to Waaigat 8. main entrance to Wilhelmina 9. north entrance to motorized traffic

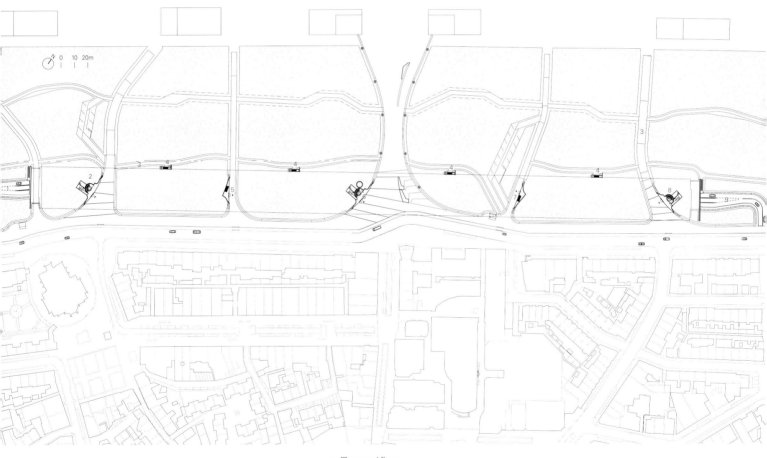

一层 ground floor

A-A' 剖面图 section A-A'

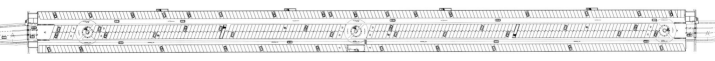

地下一层 first floor below ground

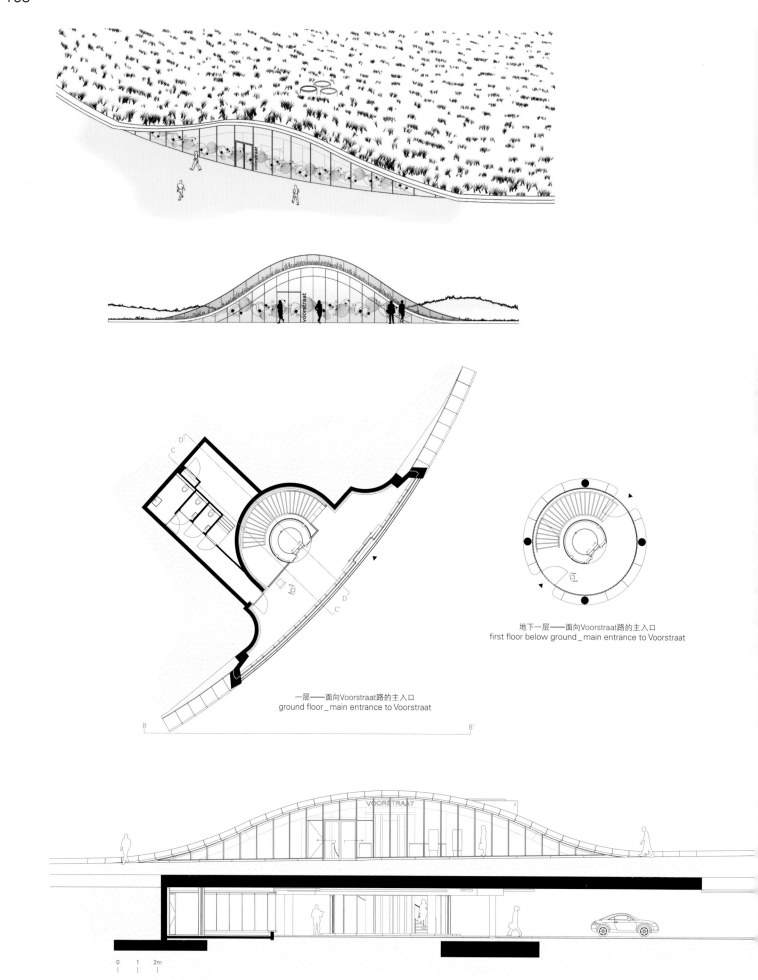

一层——面向Voorstraat路的主入口
ground floor _ main entrance to Voorstraat

地下一层——面向Voorstraat路的主入口
first floor below ground _ main entrance to Voorstraat

B-B'剖面图——面向Voorstraat路的主入口
section B-B' _ main entrance to Voorstraat

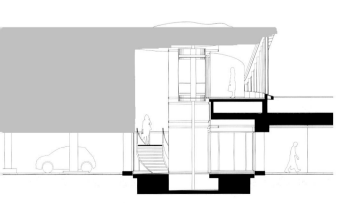

C-C'剖面图——面向Voorstraat路的主入口
section C-C'_main entrance to Voorstraat

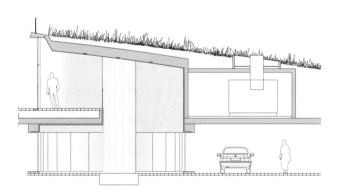

D-D'剖面图——面向Voorstraat路的主入口
section D-D'_main entrance to Voorstraat

项目名称：Underground Parking in Katwijk aan Zee / 地点：Katwijk aan zee, province of South Holland, The Netherlands
事务所：Royal HaskoningDHV - Richard van den Brule, Filipa Vieira Santos / 景观建筑师：OKRA Landscape architects - Bart Dijk
合同管理：WB de Ruimte / 沙丘顾问：Arcadis / 承包商：Bouwcombinatie Ballast Nedam - Rohde Nielsen
结构设计：Adviesbureau Snijders / 机电管道设计：De Bosman Bedrijven / 电梯设施：Liften- en Machinefabriek Lakeman
立面工程与实现：Metadecor / 整体设计（工程与施工合同）：Ballast Nedam Engineering in cooperation with Zwarts & Jansma Architects (ZJA) / 功能：mixed use, infrastructure, gardens & parks, parking, public space, waterfront /用地面积：1,800,000m²
总楼面积：15,000m² / 客户：Municipality of Katwijk, Water Board Rijnland, Province of South Holland, Ministry of Infrastructure and Environment / 总造价：EUR 70 million / 停车场面积：1,000 €/m²
施工时间：2013—2015 / 摄影师：©Treffendbeeld - p.164~165, p.176~177; ©Jan Brantjes (courtesy of the Royal HaskoningDHV) - p.166, p.172 middle-left, p.172 bottom; ©Luuk Kramer (courtesy of the Royal HaskoningDHV) - p.169, p.170~171, p.173, p.174; ©Lakeman (courtesy of the Royal HaskoningDHV) - p.172 top, p.172 middle-right

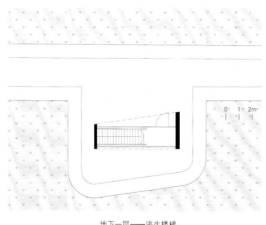

地下一层——逃生楼梯
first floor below ground _ emergency staircase

西北立面——逃生楼梯
north-west elevation _ emergency staircase

parking structure into the dune, just 3m from the dike, it became possible to increase the seaside parking capacity and to combine the two projects into one. The 663-space parking garage carefully embedded into its natural dune environment builds forth on the design of the public space and landscape. The dune has been planted with marram grass and includes paths, a boardwalk, and benches so that visitors can stroll along its crest or to the sea. These carefully shaped dunes, which rise up organically to create subtle entrances and exits ensure the character of the dunes to stay intact and allows natural daylight to flow into the underground area, benefitting the orientation inside the long, elongated parking garage. The entrances are set up identically, although the situation required a unique connection with its boulevard every time. At the seaside of the parking garage, the emergency exits are located. Minimized in size, they are designed as sculptures in the dunes. The road surface of the garage consists of interlocking bricks to have as much daylight as possible that the visitor entrances are transparent from top to bottom.

The unique setting required high quality and sustainable finishes because of its difficult weather conditions. Glass, (perforated, weatherproof) steel, porcelain tiles, and exposed concrete are the dominant materials used. To curtail nuisance to the environment, shorten the time to build the parking, and provide low maintenance, it was chosen to be built in modular with prefabricated concrete elements. These elements guaranteed a constant high quality of the wall, columns, and roof. Besides, many elements have been standardized, and identical details have been applied which strengthens its modest architectural appearance. Overall, the colors and materials used seamlessly fuse into the urban fabric of Katwijk aan Zee. Underground Parking in Katwijk aan Zee therefore successfully meets the need for defensive coastal protection, functional parking requirements as well as the desire for a landscape design which is related to its coastal environment.

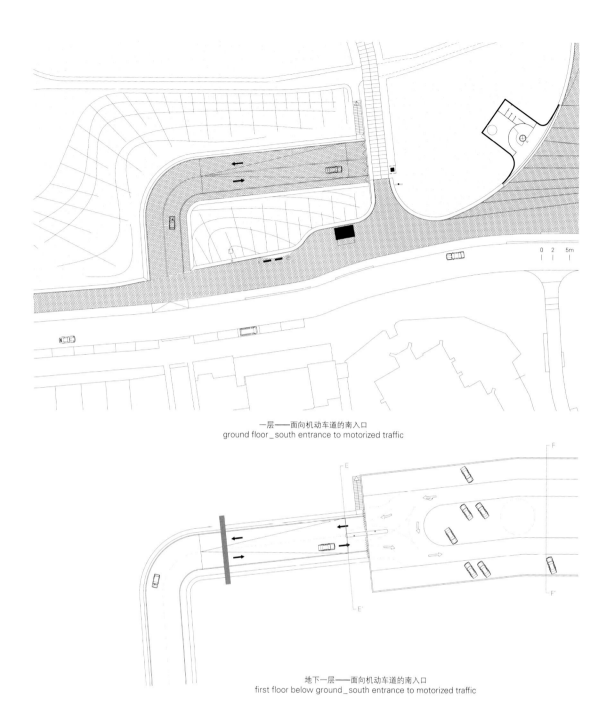

一层——面向机动车道的南入口
ground floor _ south entrance to motorized traffic

地下一层——面向机动车道的南入口
first floor below ground _ south entrance to motorized traffic

E-E'剖面图——面向机动车道的南入口
section E-E' _ south entrance to motorized traffic

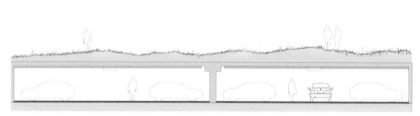

F-F'剖面图——面向机动车道的南入口
section F-F' _ south entrance to motorized traffic

里斯本游轮码头
Lisbon Cruise Terminal

Carrilho da Graça Arquitectos

里斯本位于阿尔法马山坡，眺望着塔古斯河口。山脚下的平地在20世纪初被用作港口垃圾填埋场，如今这是一座新建的游轮码头建筑——一个小型的圆形竞技场，看上去好像是背对着河岸，身子扭转着望向城市并回应着城市。这个体量紧凑的项目（曾在2010年作为国际竞赛中最小的建筑单元参赛）以及露天停车场和潮汐池夹在已经填埋的旧Jardim do Tabaco码头的墙体间。建筑位于公园/林荫大道的树木之间，仿佛脱离了地面，现已成为滨河区的延伸地带。它使公共空间从地面抬升，变成了一个露台或观景台，在河流和城市之间形成抽象地形，如同中转木筏，使二者紧紧相连，使它们相互守望。

码头的功能就被包裹在一层外壳之下：地下是停车场，与露天停车场相连；地面层用来办理行李托运、审核和赔偿等业务；上面楼层则是旅客服务区，有登记处、候船室、VIP室、免税店、公共咖啡店等。和公园/林荫大道情况一样，所有的空间都被灵活地利用，便于应对各种营运时间之外发生的意外事件，甚至可以在未来发展成航海站点。

围合建筑场地的骨架采用了带有软木的混凝土结构，这是建筑师在维持原有基础不变的限制条件下，为减轻建筑物重量专门设计的一个方案。这个方案的构想来自于建筑师自己在里斯本设计双年展上提出的实验性概念——著名的"里斯本之光"，即利用阳光在河口形成反射的光线，使建筑熠熠生辉，产生特殊的视觉品质。

从塔古斯河岸这侧来看，里斯本码头建筑似乎不太引人注意，仿佛城市里一座不起眼的石台。可是从城市这一侧望去，那些褶皱明显地表明了出入口的位置。它的设计协调了使用者与河岸和城市之间的视觉关系：这是一座几乎总是呈现动态的建筑，人们可以沿着舷梯走，通过凉廊上船，或从凉廊往下走直接进入城市，可以走在屋顶上，也可以走在通向主立面的切线方向道路上，这种视觉转换就仿佛电影镜头切换一般。

On the Alfama slope, Lisbon is looking out to the Tagus estuary. At the foot of the hill, on the flats of the early 20th century landfill of the port, the new Cruise Terminal building, a small amphitheater, apparently with its back to the river, echoes, and returns, and looks back to the city. The compact volume of the project (the smallest of the buildings in the entries presented to the international competition in 2010), is inserted – with the open-air car park and the tidal tank – between the walls of the landfilled former Jardim do Tabaco dock. Seemingly not touching the ground, the building sits between the trees of the Park/Boulevard, now inhabiting as a stretch of the river front. It lifts the public space from the ground and transforms it into a terrace/viewpoint, forming an abstract topography between the river and the city, like a transshipment raft that connects and reveals both.

The programme of the terminal is housed inside this shell: the underground car park that is connected to the open-air car park; luggage delivery, processing and claim, on the ground level; passenger areas (check-in, waiting lounge, VIP lounge, duty-free shopping, public access coffee shop) in the upper level. All spaces are flexible spaces, like those of the Park/Boulevard, that allow events of other nature to take

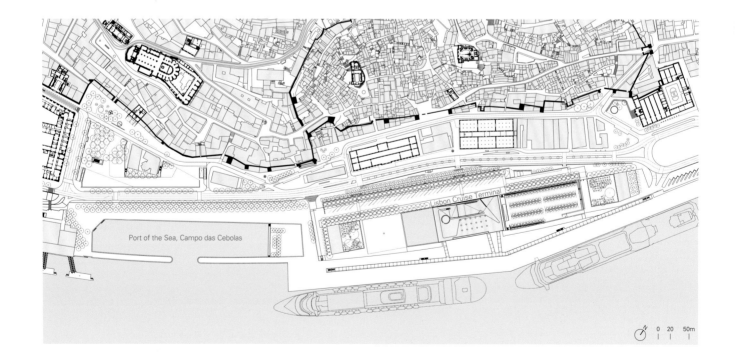

项目名称：Lisbon Cruise Terminal / 地点：Santa Apolónia, Lisbon, Portugal / 建筑师：João Luís Carrilho da Graça / 办公室经理：Francisco Freire / 项目团队：Luis Cordeiro, Nuno Pinho, Pedro Ricciardi, Paulo Costa, Yutaka Shiki, Filipe Homem, Charbel Saad, Nuno Castro Caldas, Ana Teresa Hagatong, Ana Bruto da Costa, Carlo Vincelli(architects), Nuno Pinto(draughtsman), Paulo Barreto(models) / 景观建筑设计：Global Arquitectura Paisagista Lda - João Gomes da Silva / 交流设计：P-06 atelier - Nuno Gusmão (graphic artist) / 工程协调、基础与结构、电气、电信、水气设施、安全系统：Fase-Estudos e Projectos SA / 空调、通风与消毒、能源优化与热性能、声音调节、中央技术管理：NaturalWorks-Projectos de Engenharia Lda / 海洋水力学：Consulmar-Projectistas e Consultores Lda / 环境与可持续性：Nemus-Gestão e Requalificação Ambiental Lda / 客户：Lisbon Port Authority, Lisbon Cruise Terminals / 总楼面面积：12,440m² / 总建筑造价：19.5 Mio. Euro / 竞赛时间：2010 / 规划时间：2010—2015 / 施工时间：2016—2018 / 摄影师：©FG+SG - p.178~179, p.182, p.185, p.187, p.189, p.190, p.192~193; ©FG+SG (courtesy of the architect) - p.181, p.184, p.188

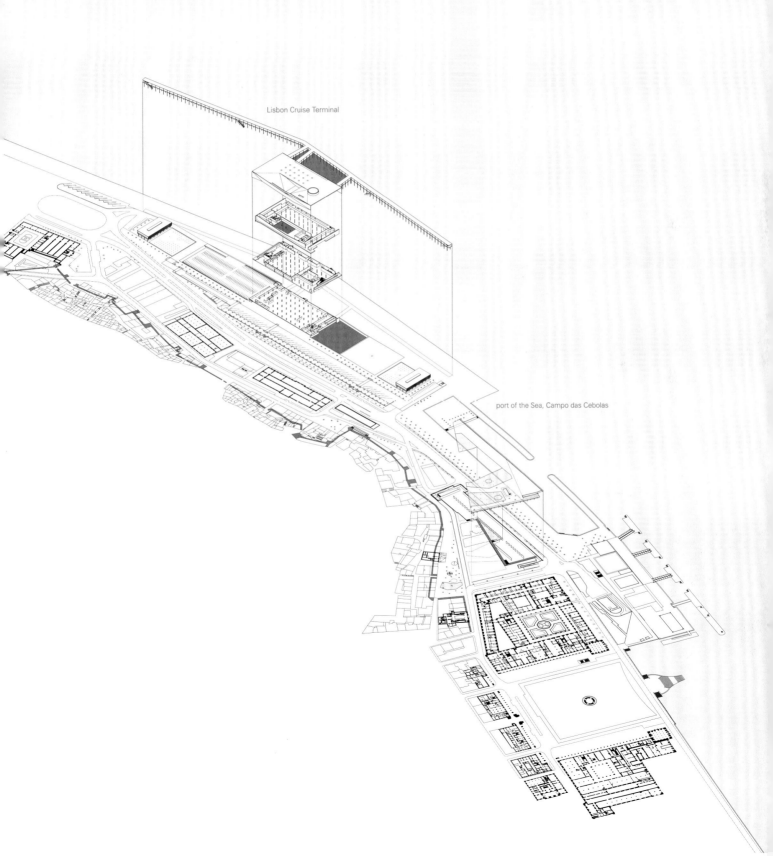

place outside the opening hours and seasons, even granting the future evolution of the maritime terminal.

This sort of exoskeleton, that encircles the areas assigned to the terminal's programme, is built with structural concrete with cork – a solution for limitations by the pre-existing foundations, specifically developed to reduce the building's weight. The idea stemmed from the architect's own concept from experimental design presented in the Lisbon design biennale – the famous "light of Lisbon" that lightens up with the sunlight reflected on the estuary, with a particular haptic quality.

Lisbon Cruise Terminal virtually seems to be blind from the river side, almost appearing as a discreet stony socle of the city. On the city side, the building creases, just enough to reveal its access points. It mediates the visual relations between its users and the river and city: in a building that is almost always in motion – along the gangway, in the loggias that give access to the ships or from these to descend directly into the city, walking on the rooftop, on the tangential approaches to the main facade – the cinematic gaze wanders.

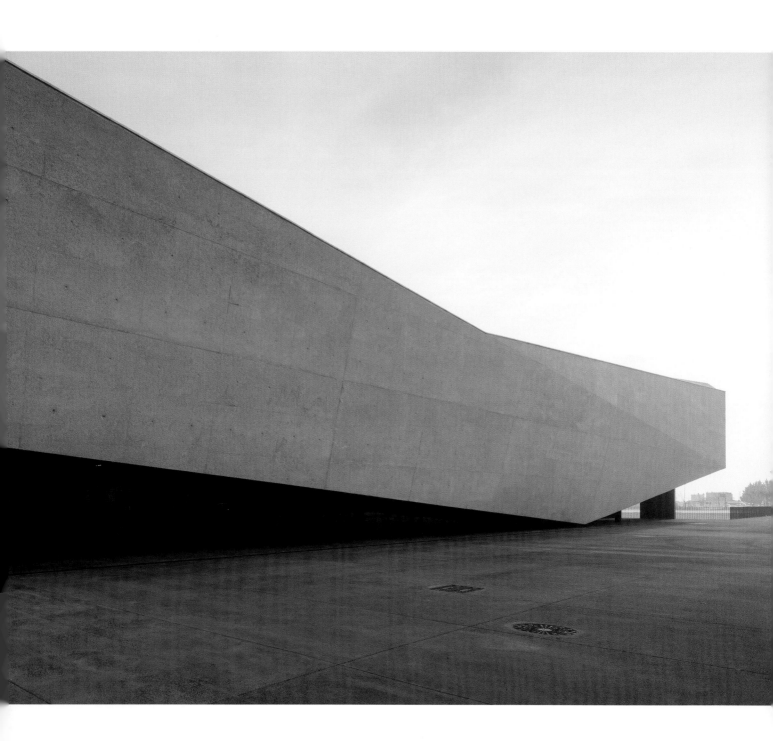

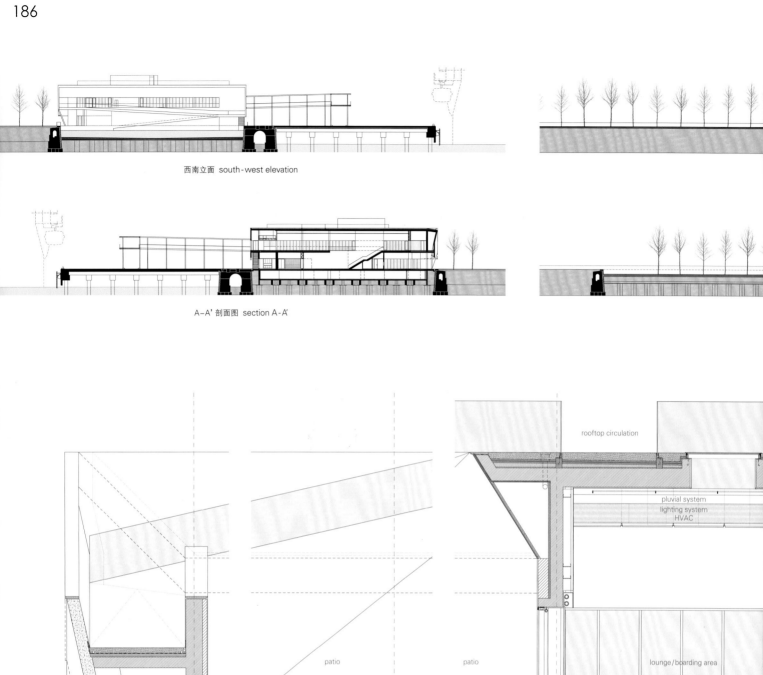

西南立面 south-west elevation

A-A' 剖面图 section A-A'

a-a' 详图 detail a-a'

西北立面 north-west elevation

B-B' 剖面图 section B-B'

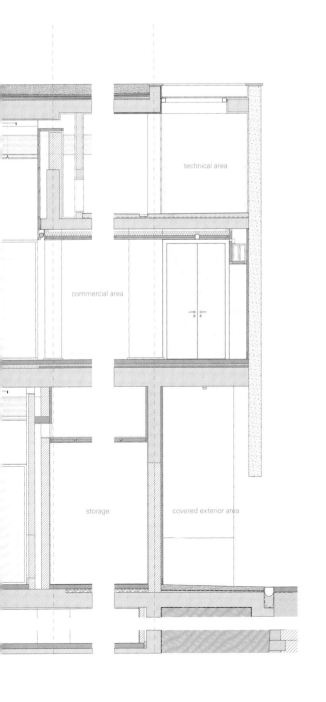

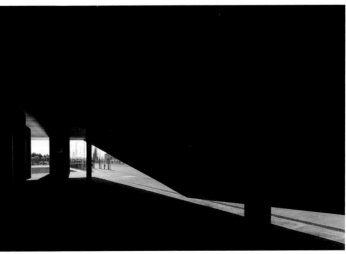

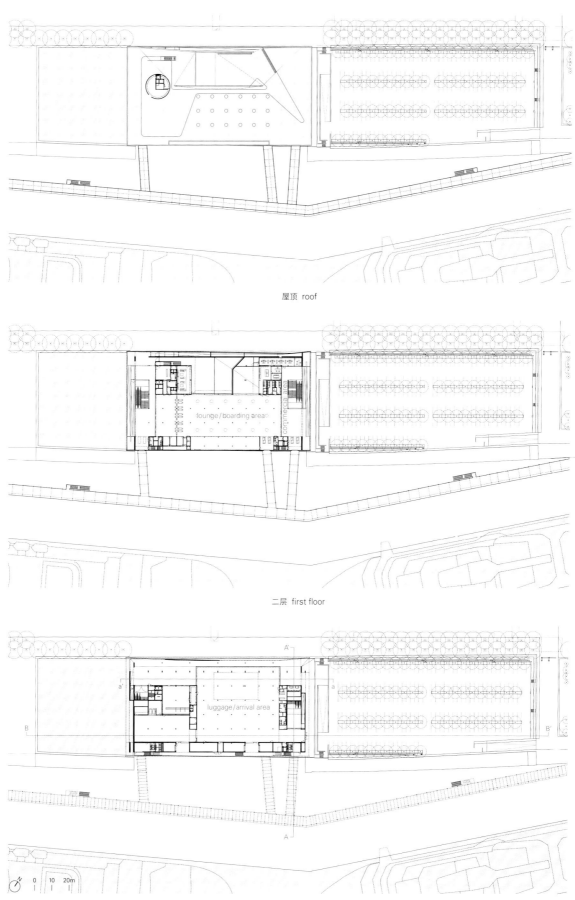

屋顶 roof

二层 first floor

一层 ground floor

波尔图游轮码头
Porto Cruise Terminal

Luís Pedro Silva, Arquitecto, Lda.

波尔图游轮码头是一个小型港口综合建筑设施。它靠近大西洋，位于波尔图马托西纽什南部码头边缘的弧形地段，距离城区只有750m，而通过河道或通过陆路，到达波尔图的里贝拉约有10km的距离，里贝拉是最受欢迎的旅游景点之一。在工程建设初始阶段，波尔图游轮码头就具有了双重战略目的：提高商业效率以及更好地促进城市一体化。项目完成后该港口就有条件接纳体积巨大，船长约为300m的巨型游轮了。而且，考虑到如今的城市居民已经不再依靠捕鱼业为生，他们与港口之间的依赖关系就好像港口跟罐头厂之间的关系一样，正在逐渐消失，而这个项目的建成会促进周边社区的社交往来。工程是分阶段进行的，依次是：一个新码头和主建筑，主建筑中心的水池，街道，滨水区，一座小型辅助建筑。

项目在很大程度上依靠主建筑，主建筑连接了三个主要的功能区：新游轮码头、滨水区以及连接城市的新街道。围绕主体建筑，大大小小的项目也开始进行，如游轮码头、滨水区设施、波尔图海洋大学科技园以及几个宴会活动厅，一间小餐厅，一个小停车场。码头的曲线形态以及连接各区域的功能因三个"叶片"缠绕式的设计而得以突显，这三个"叶片"是三个主要的室外"触角"，第四个"触角"向内下落，形成一条螺旋状的坡道，连接着四层高空间内的不同室内功能区。通过这条螺旋坡道，公众可以进入屋顶。其中三个弯曲的"触角"向不同方向延展，研究层通往海洋，出发层通往游轮舷梯，高架步行栈道通往海滩和马托西纽什市。

建筑外墙材料选用了瓷砖，它的设计灵感来源于周围的海洋环境和波尔图建筑立面材料传统的覆层。陶瓷的运用给建筑平添了个性，具有与场地内不断变化的色彩和光线和谐相融的能力。瓷砖弯曲的表面仿佛在与城市中充足的光线、海水进行对话。它的珍珠色调与黄色的花岗岩海堤产生了对比效果。除了瓷砖的纹理很有特点，金属的

围栏也因为波浪形的设计和几乎有韵律的舞蹈形式而得以突显。花岗岩也是葡萄牙北部地区一种富产的石材,被广泛应用在建筑内外的路面、台阶以及首层设计精致的服务区的墙面上。含铜的吊顶增强了建筑形态从室外到室内的过渡,抬高了这个开阔区域上方的楼层。

建筑本身并不太大,尤其是与受它保护的游轮相比,而且用的大多是寻常的建筑元素。可是,它却建立了与大海和陆地之间深刻、细致且深厚的关系。在海滩上看,它明显属于莱索斯港口,建筑面朝码头,但看不到任何一个洞口,只显露出一个神秘的封闭立面。而从远处看,它蜿蜒的白色立面却能反映出光线与大气的细微变化。靠近它,你就会被它曲线形的外观、陶瓷的纹理和灵动的体态所吸引,想要看一看,摸一摸。建筑师在设计时兼顾景观与建筑,精心打造出包含多种功能的综合建筑,使波尔图游轮码头成为这个城市和葡萄牙北部地区公共空间的重要设计标杆。

Porto Cruise Terminal is a small port complex located at the end of the south jetty in its curved sector, within the Atlantic Ocean, 750m away from Matosinhos city and 10km away via river-sea or road from Porto Ribeira, one of the most popular touristic point. Since the beginning, the project has a double objective: an improvement of the commercial efficiency and a better urban integration. It aims to receive large cruise ships, up to 300m, and integrate spaces with an urban vocation to encourage sociability among the surrounding urban community, given that nowadays the dependency bond between the city's population and the fishing activity is dissipated such as that between the port

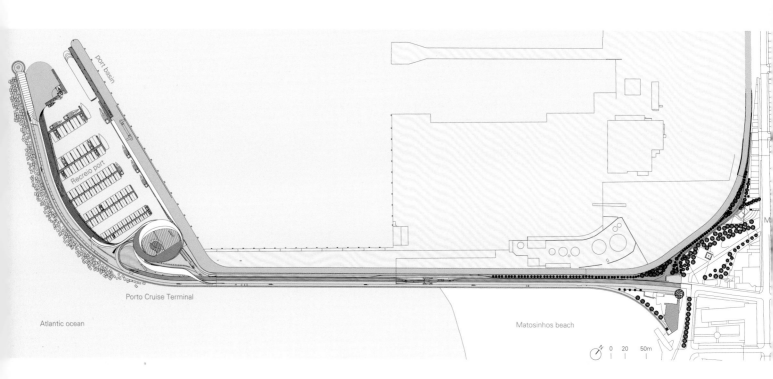

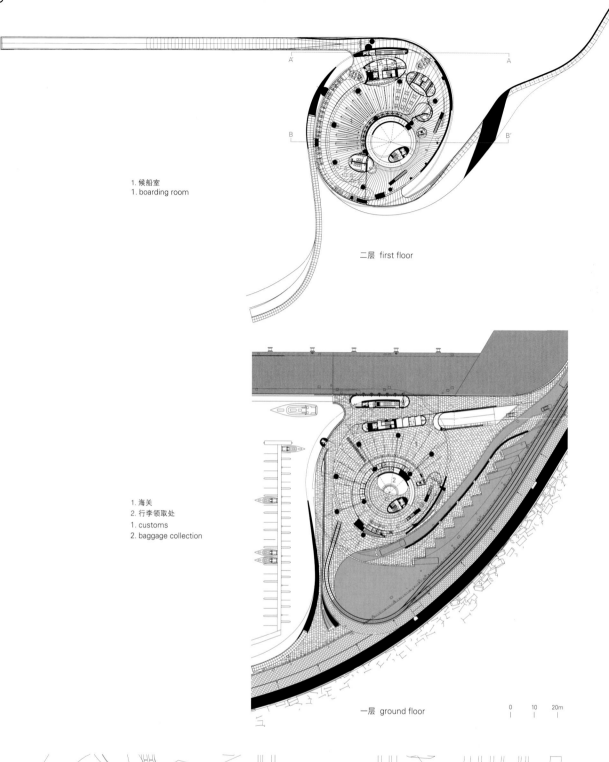

1. 候船室
1. boarding room

二层 first floor

1. 海关
2. 行李领取处
1. customs
2. baggage collection

一层 ground floor

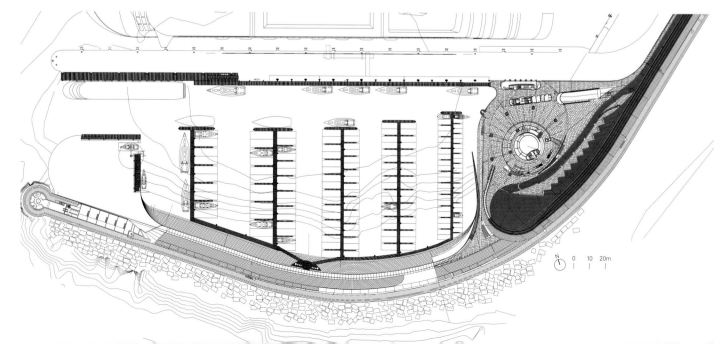

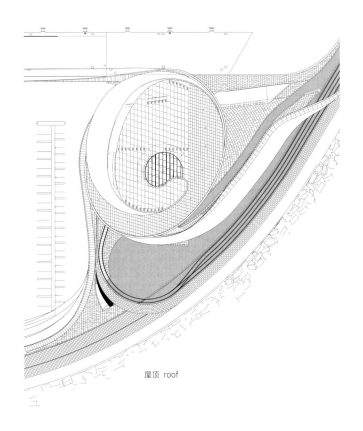

屋顶 roof

1. 科学宣传室
2. 多功能室
3. 餐厅
4. 屋顶露台

1. science dissemination room
2. multipurpose room
3. restaurant
4. roof terrace

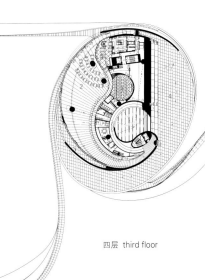

四层 third floor

1. 机械实验室
2. 机器人实验室1
3. 机器人实验室2
4. 生物多样性实验室
5. 生态学实验室
6. 应用微生物实验室
7. 生态生理学实验室
8. 免疫生物学实验室
9. 营养、生长与品质实验室
10. 环境毒理学实验室
11. 细胞、分子与分析研究实验室
12. 化学与常用设备实验室
13. Augusto Nobre毒物学实验室
14. 生态毒理学实验室

1. equipment laboratory
2. robotics 1 laboratory
3. robotics 2 laboratory
4. biodiversity laboratory
5. ecology laboratory
6. applied microbiology laboratory
7. ecophysiology laboratory
8. immunobiology laboratory
9. nutrition, growth and quality laboratory
10. environmental toxicology laboratory
11. cell, molecular and analytical studies laboratory
12. chemistry and common equipment laboratory
13. Augusto Nobre toxicology laboratory
14. ecotoxicology laboratory

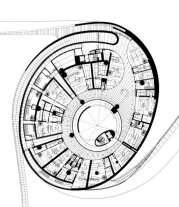

三层 second floor

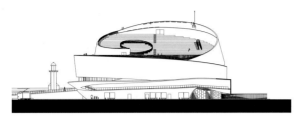

东南立面 south-east elevation

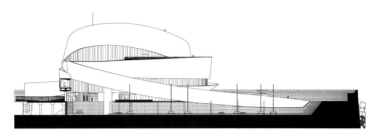

西北立面 north-west elevation

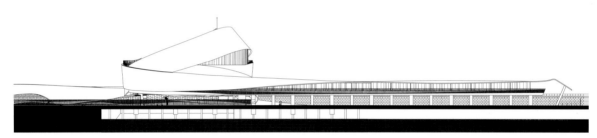

东北立面 north-east elevation

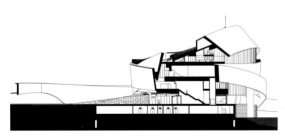

A-A' 剖面图 section A-A'

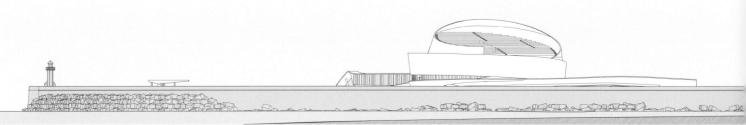

南立面 south elevation

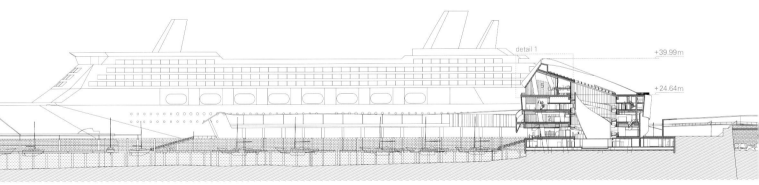

1. 行李领取处 2. 候船室 3. 细胞、分子与分析研究实验室 4. 生物多样性实验室 5. 多功能室
1. baggage collection 2. boarding room 3. cell, molecular and analytical studies laboratory 4. biodiversity laboratory 5. multipurpose room
B-B' 剖面图 section B-B'

项目名称：Porto Cruise Terminal
地点：south pier of Leixões port, Matosinhos, Porto, Portugal
事务所：Luís Pedro Silva, Arquitecto. Lda
结构工程师：NEWTON, Consultores de Engenharia, Lda
电气工程师：Rodrigues Gomes & Associados – Consultores de Engenharia, S.A
机械工程师：GM Engenharia
室外布局：José Magalhães
总承包商：ACE
客户：APDL (Administração dos Portos do Douro, Leixões e Viana do Castelo), Oporto University
总楼面面积：17,500m² (main building)
供应商：Granite – Transgranitos; Precast concrete – Pavicentro; Ceramic tiles – Vista Alegre; Windows and railings – Jofebar, Metaloviana; Glazing – Jofebar, VCP; Furnishings – Laborial (Laboratory furniture), Luís Pedro Silva, Arquitecto. Lda (Laboratory Hottes design), Laborial (supply); Lighting design – Luís Pedro Silva, Arquitecto. Lda; Lighting supply – Osvaldo Matos O/M, Ramos Ferreira, Exporlux
竣工时间：2015.3
摄影师：©FG+SG

and the canneries. A new quay and the main building, the reflecting pool at the core of the main building, the street, the marina settlement, and the setup of a small support building were therefore completed in stages.

Yet, the project is largely dependent on the main building, which constitutes a node among the three main functions: the new cruise ship quay, the new marina, and the new street linking to the city. It magnetizes several programmatic components of cruise ship terminal, marina facilities, the Science and Technology Park of the Sea of the University of Porto, event rooms, a small restaurant, and a parking lot. Enchanted by the jetty's curve and this intermediary commitment, it enlaces the curved blades in the form of three main exterior tentacles and the fourth falling inwards to a helical ramp connecting the internal functions within a quadruple height space. Through the helical ramp, the public can further access the rooftop. The unrolled exterior arms lead the

investigators level to the sea-side and the departure level to the cruise gangway, the elevated walkway towards the beach and Matosinhos city.

The building coated in ceramic tiles is inspired by the maritime environment and the traditional coating of Porto's building facades. They add on to the building's personality, conferring its capacity to harmonize with the chromatic and light variations of the site; the curved surfaces with the ceramic tiles speak with the abundant sunlight and water in the city. Its pearl tone also contrasts with the yellow granite of the port's seawall. Besides the texture of the tiles, the metalwork for the rail is highlighted by its wavy progression, almost rhythmically dancing. The granite, another abundant material in Northern Portugal region, was used for the interior and exterior pavements, the steps, the wall finish of the elaborated level 0 service volumes. A coppery suspended ceiling emphasizes the progression from the exterior to the interior, lifting the upper levels from this broad area.

The building is not very large, mostly when compared with the cruise ships it can shelter and is motivated by prosaic elements. However, it builds up thoughtful, meticulous, and sedimentary relation with the sea and the land. From the beach side, it clearly belongs to Leixões port, in such a manner that it faces towards the jetty, not revealing any openings and declaring a mysterious blind facade. From far away, its sinuous white facade is read with varying nuances of light and atmosphere. The arches and its texture appeal to the proximity, involve movements and the body, inviting the look and the touch. Designed simultaneously for the landscape and the body, deliberately thought to acquire the complex of different functions, Porto Cruise Terminal is now a reference for the city's and Northern Portugal's public space.

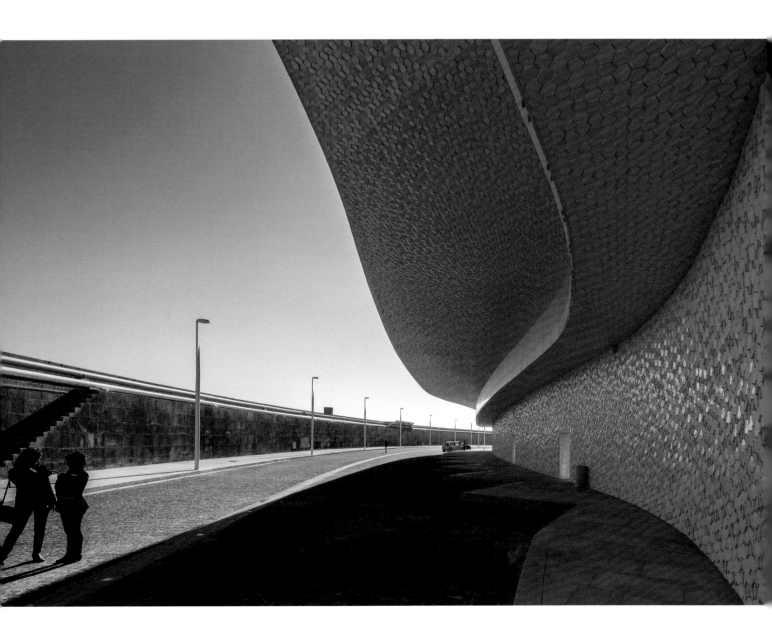

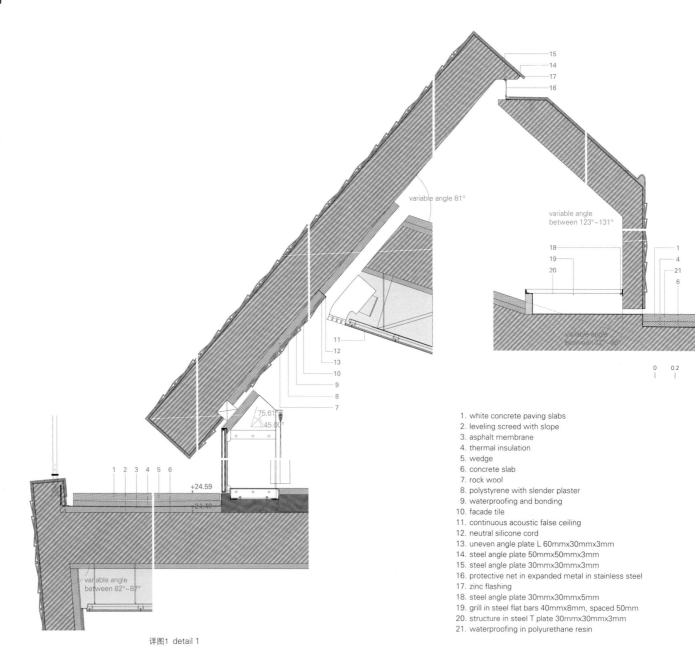

1. white concrete paving slabs
2. leveling screed with slope
3. asphalt membrane
4. thermal insulation
5. wedge
6. concrete slab
7. rock wool
8. polystyrene with slender plaster
9. waterproofing and bonding
10. facade tile
11. continuous acoustic false ceiling
12. neutral silicone cord
13. uneven angle plate L 60mmx30mmx3mm
14. steel angle plate 50mmx50mmx3mm
15. steel angle plate 30mmx30mmx3mm
16. protective net in expanded metal in stainless steel
17. zinc flashing
18. steel angle plate 30mmx30mmx5mm
19. grill in steel flat bars 40mmx8mm, spaced 50mm
20. structure in steel T plate 30mmx30mmx3mm
21. waterproofing in polyurethane resin

详图1 detail 1

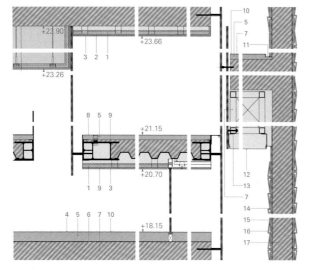

1. rock wool
2. waterproof plasterboard
3. paint with anti-fungus additive
4. vinyl
5. screed improved with special cement
6. separation blade with 5mm thickness
7. concrete slab
8. linoleum
9. corrugated concrete slab
10. cementitious pavement 2 to 4mm
11. stainless steel angle plate 60mmx30mmx3mm
12. tensioned canvas
13. metallic structure
14. facade tiles
15. bonding
16. waterproofing
17. grey concrete blade

详图2 detail 2

西港 2 号码头大楼
West Terminal 2

PES-Architects

西港2号码头大楼坐落于赫尔辛基西港口一片狭窄的填海地带上，处于新建住宅区的最南端。新码头大楼是在赫尔辛基—塔林航线轮渡交通需求日益增长的背景下建成的，其目的是提高旅客登船和下船的速度，并将渡船在港口内的轮班时间缩短至一个小时。

码头大楼建在两个船埠之间，充分利用狭小的场地。由于船岸大部分的区域要用来停列等候登轮的车辆，因此首层的空间设计上做到尽可能地紧凑。二层的离港大厅与地面相距10m，使得建筑下方的交通更加顺畅。设计师把大厅定位在码头中心，并将其位置抬高，使旅客们的步行距离被缩到最短，从而使他们更迅速地经由桥和走廊登上码头任何一侧的船只。

码头大楼首层主要用作通行区域，设有多个出入大厅。旅客先在围有15m高的玻璃墙的检票区进行检票，进入大楼之后，乘坐扶梯或直梯迅速地来到离港大厅。离港大厅的空间宽阔，像一个飞机库，里边包含餐饮区、咖啡厅和就座区。透过高高的玻璃墙，可以欣赏到辽阔的海面与来往的船只。大厅上方的大型天花板距离地面9m，由经过热处理的松木板条构成，其中容纳了照明、通风和消防喷淋装置。天花板由中间向两侧缓缓降低，把旅客引向登船桥的方向。玻璃墙面和照明设备共同保证旅客通行路线的清晰与安全。离港和到港旅客的交通路径彼此分离，到港旅客直接被引导至扶梯处，穿过一层海关大厅走出码头。码头大楼框架的几何形状十分复杂，尤其是从屋脊处开始双向弯曲的屋顶，这一形状是采用钢框架结构实现的。

除了满足高效性之外，设计师还使码头大楼成为品质一流、环境舒适、造型优美的公共设施。这座码头大楼在Jätkäsaari开发区俨然已成为地标式公共建筑。建造中使用的抗磨蚀材料能够保持码头大楼在未来至少几百年间魅力不减。

从外部看，光滑流畅的轮廓让建筑看上去就像被冲上岸的海洋生物，玻璃、混凝土和铝制成的立面在阳光下像海面一样波光粼粼。主入口附近的松木墙面追忆着用木头造船的年代。

公共交通使旅客来往码头大楼更加便捷，出租车站点就位于主入口旁边，配有由钢材和玻璃制作的大型雨篷，进一步使西港2号码头大楼成为这座城市未来的公共交通枢纽。

West Terminal 2 is a new ferry terminal situated on a narrow plot in the port expansion area at the southern tip of the new housing district in Helsinki's West Harbor. The new terminal was built to meet the needs of the growing ferry

traffic on the Helsinki-Tallinn route. The goal was to enable faster embarkation and disembarkation of passengers and reduce the turnaround times of ferries in port to just one hour.

The terminal building is located between two quays to make the most of the relatively small plot. As most of the dock area is required for the vehicles queuing to board the ferries, the ground level was designed to be as compact as possible. The second-level departure lounge was raised 10 meters off the ground, allowing traffic to flow smoothly under the building. The central, raised location of the lounge also minimises passengers' walking distances along the corridors and bridges to the ships on either side of the terminal.

The ground floor is mainly a pass-through area with separate exit and entrance lobbies. Passengers enter the terminal through a check-in area with glass walls 15 meters

项目名称：West Terminal 2 /地点：Tyynenmerenkatu 14, Helsinki, Finland / 事务所：Tuomas Silvennoinen (Partner), Pekka Mäkelä (Project architect), Emanuel Lopes, Hanna Eskelinen / 项目团队：Onni Aho, Milla Alaraatikka, Aino Aropaltio, Sam Cowley, Riku Haaramäki, Tiina Juntunen, Mikko Karppanen, Sami Lauritsalo, Kai Lindvall, Toivo Moustgaard, André van Tulder / 结构工程师：Sweco Rakennetekniikka Oy / 暖通空调工程师：Ramboll Finland Oy / 电气工程师：Granlund Oy
开发商顾问：Viator Oy, Indepro Oy / 总承包商：YIT Rakennus Oy / 电气承包商：LSK Talotekniikka Oy / 空调承包商：Kaunisto-Yhtiöt / 管道承包商：Star Expert Oy
客户：Port of Helsinki Ltd / 总楼面面积：12,900m² / 材料：Exterior, facade - glass, aluminum sheets (6mm, powder coated), concrete element, thermo treated pine (next to entrance); Exterior, roof - standing seam roof (aluminum, falzonal); Interior, walls - birch tree paneling, aluminum sheets (4mm, powder coated), concrete, glass; Interior, ceilings – aluminum (anodized, 3mm, entrance/exit hall), thermo treated pine batten / 设计开始时间：2014 / 竣工时间：2017 / 摄影师：©Kari Palsila (courtesy of the architect) - p.216, p.218, p.219, p.224~225, p.226; ©Marc Goodwin (courtesy of the architect) - p.214~215, p.218, p.222~223, p.227

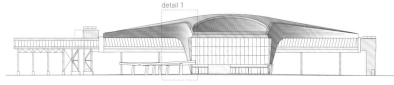

东北立面 north-east elevation

西南立面 south-west elevation

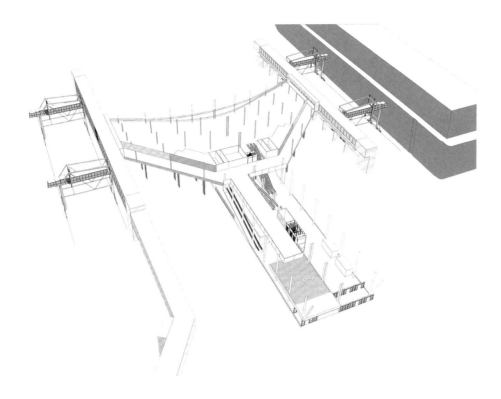

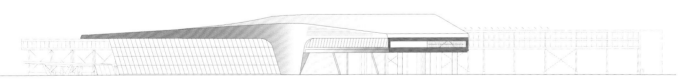

西北立面 north-west elevation

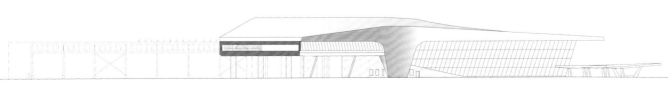

东南立面 south-east elevation

0 5 10m

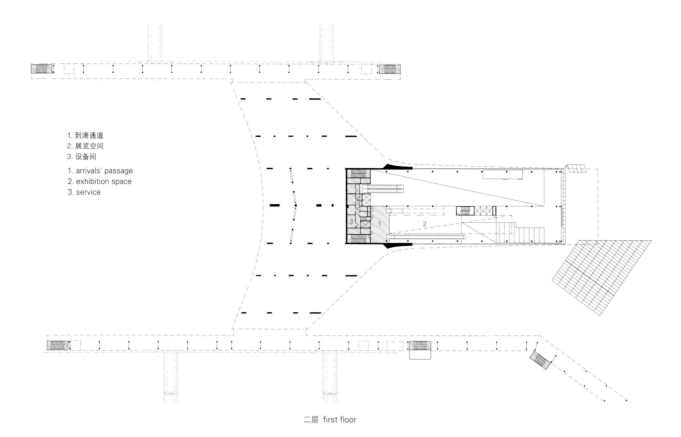

1. 到港通道
2. 展览空间
3. 设备间

1. arrivals' passage
2. exhibition space
3. service

二层 first floor

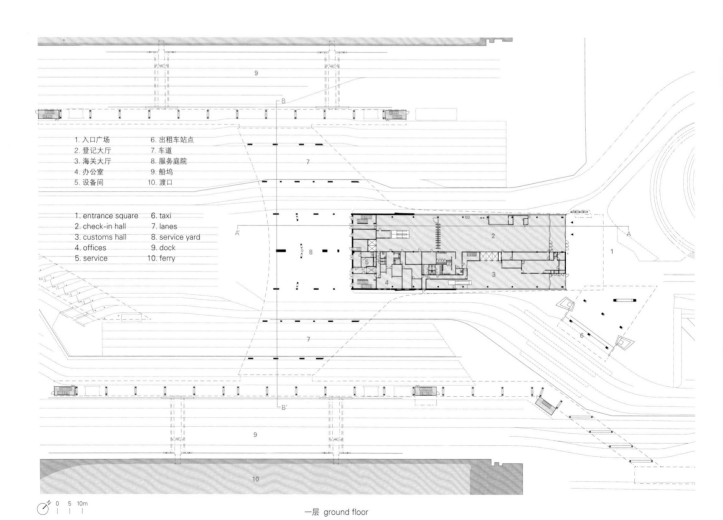

1. 入口广场　　6. 出租车站点
2. 登记大厅　　7. 车道
3. 海关大厅　　8. 服务庭院
4. 办公室　　　9. 船坞
5. 设备间　　　10. 渡口

1. entrance square　6. taxi
2. check-in hall　　7. lanes
3. customs hall　　 8. service yard
4. offices　　　　　9. dock
5. service　　　　　10. ferry

一层 ground floor

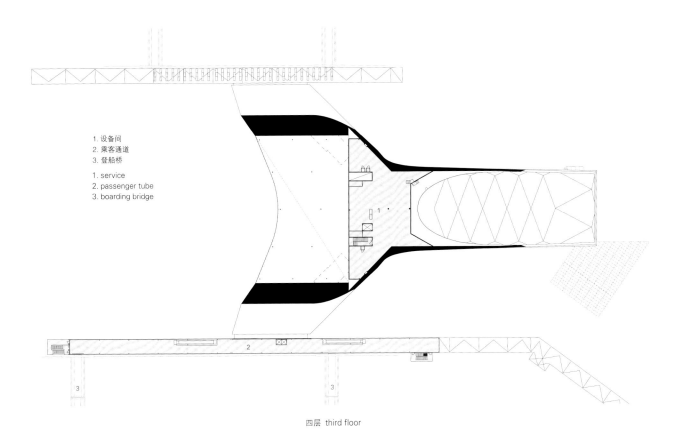

1. 设备间
2. 乘客通道
3. 登船桥

1. service
2. passenger tube
3. boarding bridge

四层 third floor

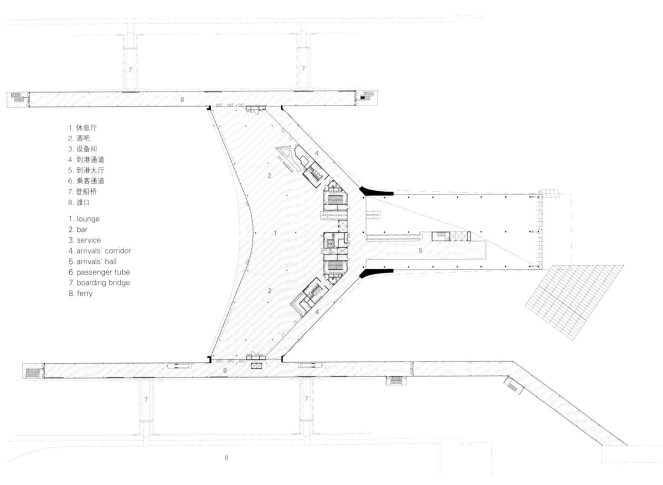

1. 休息厅
2. 酒吧
3. 设备间
4. 到港通道
5. 到港大厅
6. 乘客通道
7. 登船桥
8. 渡口

1. lounge
2. bar
3. service
4. arrivals' corridor
5. arrivals' hall
6. passenger tube
7. boarding bridge
8. ferry

三层 second floor

high and then quickly progress to the departure lounge on escalators and lifts. The departure lounge is a spacious, hangar-like space with a restaurant, café, and seating. High glass walls open to a sweeping view of the sea and incoming and departing ships. The expansive wooden ceiling, made from heat-treated pine slats, rises up to a height of nine meters. Lighting, ventilation, and sprinklers are integrated discretely into the ceiling. From the middle, the ceiling slopes down towards the sides, directing passengers to the boarding bridges. Glass surfaces and lighting solutions play an important role in keeping the passenger route clear and safe. The flows of departing and arriving passengers are separated, with arriving passengers being guided directly onto escalators and out of the terminal through customs on the ground floor. The geometry of the terminal frame is complex, particularly the roof that curves bi-directionally from the ridge. This was implemented with a steel frame structure.

In addition to the requirement for efficiency, the design aimed at a high standard of elegance, quality, and comfort. The terminal is a landmark and a significant public building in the developing urban environment of Jätkäsaari. Therefore, the materials were chosen for wear-resistance so as to retain their attractiveness throughout the lifespan of the building, at least for the next hundred years.

From outside, the sleek, flowing lines of the building resemble a sea creature washed ashore, with glass, concrete, and sea aluminum facades that gleam in the sun. Near the main entrance, the pine boards of the facade commemorate the time when ships were still made of wood.

The terminal is well served by public transportation. Next to the main entrance, a large canopy of steel and glass shelters a taxi station, further promoting West Terminal 2 as an upcoming public transportation hub in the city.

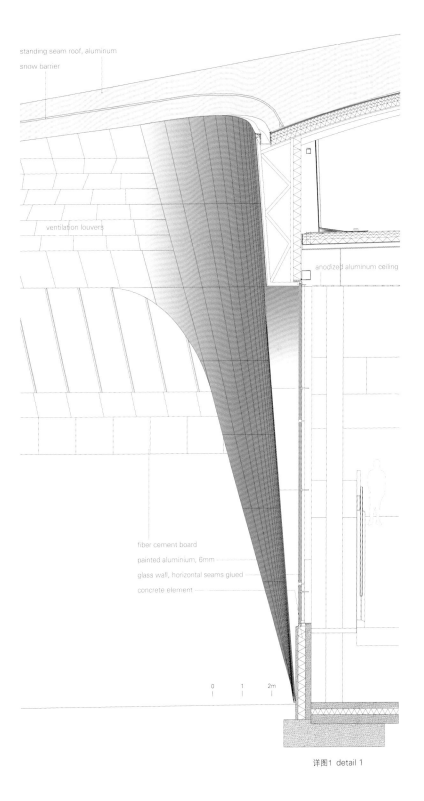

详图1 detail 1

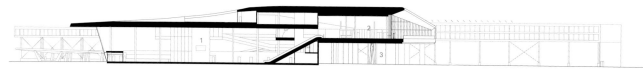

1. 登记大厅 2. 休息厅 3. 服务庭院　1. check-in hall　2. lounge　3. service yard
A-A' 剖面图　section A-A'

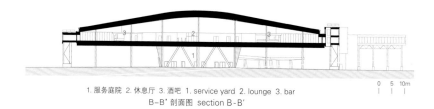

1. 服务庭院 2. 休息厅 3. 酒吧　1. service yard　2. lounge　3. bar
B-B' 剖面图　section B-B'

克罗顿滤水厂
Croton Water Filtration Plant

Grimshaw Architects

纽约市发起了"建筑设计与卓越构造倡议",旨在提升公共基础设施的设计,而位于布朗克斯区的克罗顿滤水厂则是其中一个可持续的前沿项目。该项目地处历史悠久的范歌兰公园,这里是多家机构共同所有的敏感区域。其建设需要在当地现有的和改造后的各种条件之间达到微妙而谨慎的平衡,减少彼此间的摩擦,这些条件包括实际的规划功能、安全、水文和生态系统、建筑材料的选择和美学追求。项目除了滤水厂,还包括一个约4.5ha的高尔夫球场、一家高尔夫俱乐部、几座发球台和几个停车场。项目在用途上和结构上都显现出复杂性,这也意味着该项目是一个拥有多个功能的系统,每个系统都对其他系统有直接或间接的影响。剩下部分作为二期项目已经在格雷姆肖建筑师事务所的监管下开始施工,包括公园部门设施,例如,Mosholu高尔夫球场会所、第一发球台和发球区结构。未来一期和二期项目的完成将为建筑和景观的一体化设计带来曙光。

该项目的基本宗旨就是把水这个保证人类健康的重要资源,作为安全、景观和建筑的核心要素。雨水和地下水通过景观的干预和场地吸收系统被收集起来并重新分配。在这一环节上设计师参考了睡莲的形状以及这种植物天然的滤水过程。睡莲收集雨水,过滤后供自己使用,并将多余的水送回下面的池塘,整个收集—再分配的循环过程由重力作用维持。由此,设计师通过使用生态湿地和天然沟渠,使水直接汇入收集池过滤,而无须使用人工的水泵、管道或阀门。

这些"护城河"还可以作为必要的安全边界,保护植物,因此无须再安装遮挡视线的栅栏。这是一个多功能的设计策略,不但提供了栖息地和各种设施,还整修了公园,增强其视觉效果,最重要的是保证了场地的安全。经过过滤的雨水又被重新利用给该地区灌溉,项目目标是每天处理和输送约132万吨的水——占该城市供水量的30%。

地下滤水厂的屋顶是一个3.6ha的不透水屋面,原本会导致雨水从本该入渗的地方流失。为了防止这种后果发生,该项目组称,他们将借此机会创造一个绵延起伏、有安全保障的屋面,为新球场、"护城河"和生态湿地服务。他们把约176 000m³的轻质工程土与地工泡棉结合起来,最大限度地减轻了地下工厂的结构负担。构成的地形也掩盖了工厂的空气处理机械设备,并与圆形结构整合到一起。

建筑结构在材料选用上也很独到。总体积达6880m³的石头砌成了一道全长1010m的景观护墙,把整个场地包围起来。石墙的铺设贴合了顶部的坡度,仿佛是从公园中生长出来的。耐候钢板墙由376块面板组

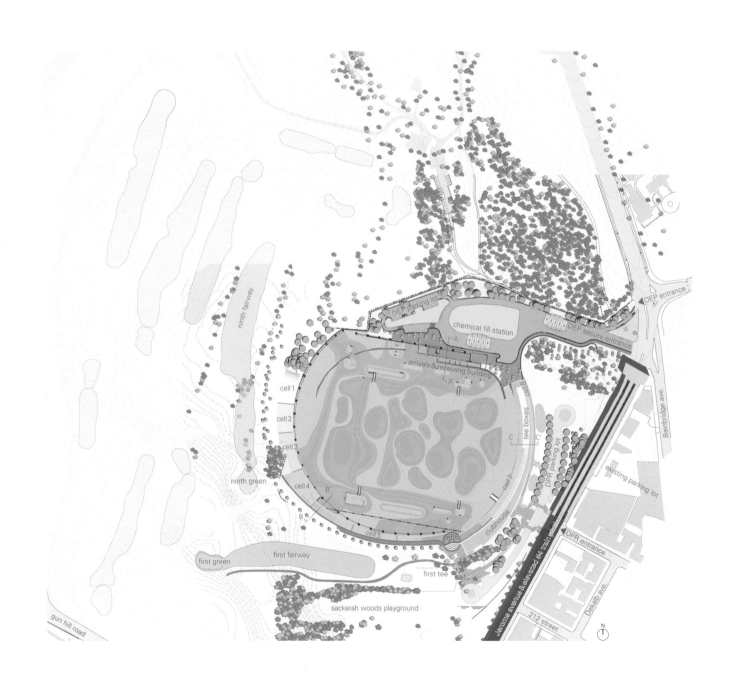

成，向外面的景观方向延伸61m，保护着迎宾大楼。为了获得半透明的效果，设计师在墙体材料上采用了穿孔设计，越远离建筑墙体越细，最后消失在周围景观中。迎宾大楼是用建筑用混凝土建造的，表现出极佳的性能，既有能承担强压力的密集配筋，又有着高质量的表层饰面。所有的建筑在外观上都非常统一，而且绿色屋顶成为高尔夫俱乐部中使用的木柱的补充。

The Croton Water Filtration Plant in the Bronx, New York, embodies emerging sustainable practices at the forefront of public infrastructure design under the New York City Design and Construction Excellence Initiative (D+CE). Located in an environmentally sensitive site of historic Van Cortland Park, the project mitigated a delicate and deliberate balance between new and existing conditions of various programmatic functions, security concerns, hydrological and ecological systems, and aesthetic/material goals within a multi-agency owned site. Beside water filtration plant, it also incorporates a 4.5ha driving range, a clubhouse, tee boxes, and parking lots. The complexity of programmatic and structural integration suggests a system that serves multiple functions, each of which directly and indirectly influences others. The remaining project construction, Phase 2, which includes the park's department facilities such as the Mosholu Golf Course clubhouse, first tee, and tee box structures, has begun under construction administration services provided by Grimshaw Architects. When the project completes with finalized Phase 1 and Phase 2, it will shine a light on the integrated design of buildings and landscape.

The project's fundamental premise, water, a vital resource to the health of our population, has been used as the generat-

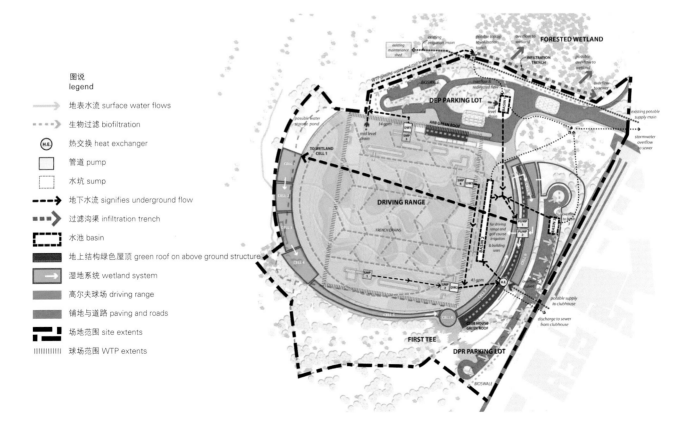

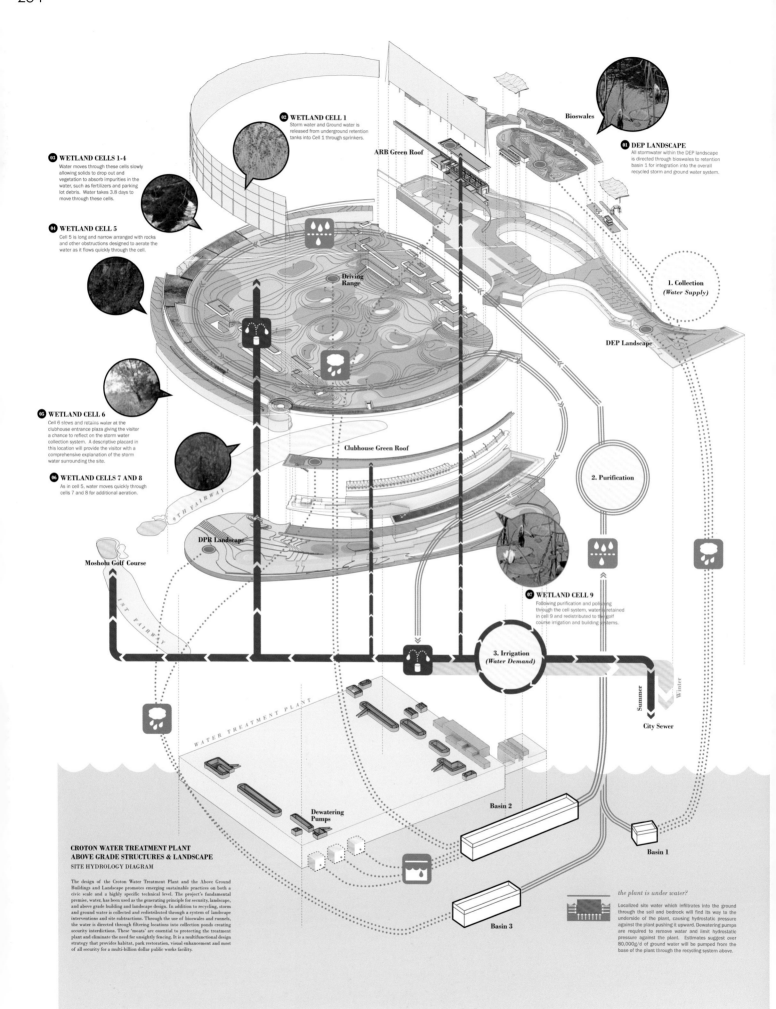

1. 绿色屋顶	1. green roof
2. 天窗玻璃	2. clerestory glazing
3. 蓝色石材覆层	3. blue stone cladding
4. 预制混凝土板覆层	4. precast concrete panel cladding
5. 维修通道	5. catwalk
6. 支撑结构	6. support structure
7. 穿孔耐候钢墙板	7. perforated weathering steel wall panel

A-A' 剖面图 section A-A'

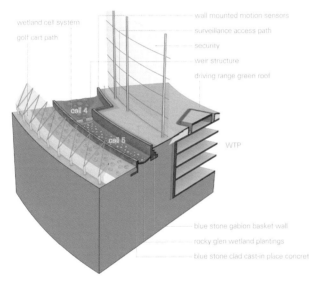

B-B' 剖面图 section B-B'

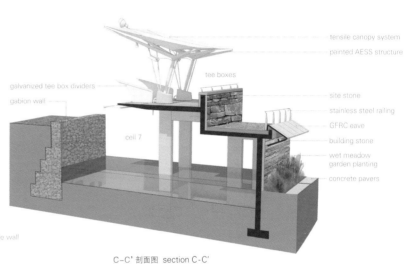

C-C' 剖面图 section C-C'

©Phillip Kuehne (courtesy of the architect)

项目名称：Croton Water Filtration Plant
地点：Bronx, New York, USA
事务所：Grimshaw Architects
结构与土木工程师：Ammann & Whitney
机电管道工程师：Buro Happold / 顾问：Ken Smith Landscape Architect (Landscape), Arup (Lighting), Rana Creek (Green roof), Northern Design (Irrigation), Atelier Ten (Sustainability) / 总承包商：Skanska Civil
客户：New York City Department of Environmental Protection
场地面积（包括水处理厂屋顶和高尔夫球场）：53,045m²
俱乐部建筑面积：1,311.23m² / 俱乐部总楼面积：1,185.16m²
结构：concrete - buildings; steel - canopy beams, net poles; Weathering steel wall; Mariani metals
室外饰面：Masonry - New York Quarries (fabricator); Berardi stone setting (installer); Metal panels - Mariani Metals (fabricator); Metal/glass curtain wall - Norshield (fabricator); Precast concrete - Coreslab Structures (exterior precast panel & canopy column fabricator); Built-up roofing - Hydrotech (WP)
室内饰面：Window metal frame - Kawneer; Glass glazing - Viracon, Cristacurva; Interior ambient lighting - Elliptipar; Downlights - Se'lux, Lucifer, Lightolier; Exterior lighting - Bega, B-K Lighting, Kim Lighting
Conveyance: Elevators/Escalators - Heights Elevator Corp.; Accessibility provision (lifts, ramping, etc.) - Advance Lifts
设计时间：2009 / 施工时间：2011—2121
摄影师：©Alex MacLean (courtesy of the architect) (except as noted)

ing principle for security, landscape, and building. Storm and ground water is collected and redistributed through a system of landscape interventions and site subtractions. Taking reference from the shape and natural filtration process of the water lily, which catches rainwater, filters it for its own use and returns the excess to the pond below, the plant's circular footprint collects and redistributes ground water through a process that is predominantly gravity-fed. Through the use of bioswales and runnels, the water is directed into collection ponds and filtering locations without the use of pumps, pipes or valves. These "moats" also serve as security boundaries necessary to protect the plant and eliminate the need for unsightly fencing. It is a multifunctional design strategy that provides habitat, park restoration, visual enhancement, facilities, and above all, site security.

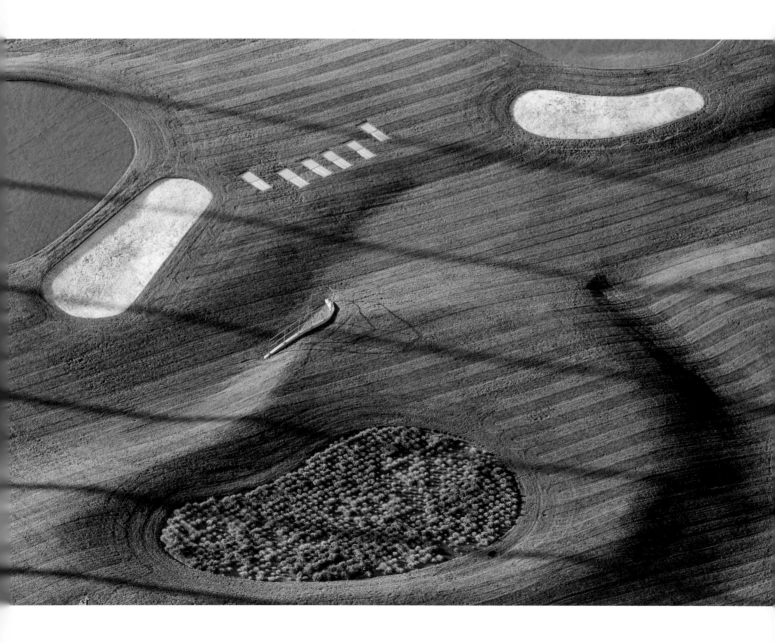

Upon filtration, the stormwater is eventually reused to irrigate the site. It is designed to treat and deliver 1.32m tons of water per day – up to 30% of the city's water supply. Located over the subterranean water filtration plant, the plant's roof, a 3.6ha impermeable surface, would cause rainwater to flow off the site where soil infiltration once occurred. To prevent this impact, the project claims the roof as an opportunity to provide a rolling, secure green surface for the new driving range, the "moat" system, and habitat creation. Approximately 176.000m³ lightweight engineered soil combined with geofoam minimises the structural burden on the plant below. The topography also masks the plant's air handling machinery and integrates the circular structures.

The building construction at Croton includes several unique material uses. 1010 meters of stone-clad landscape retaining walls from 6880m³ stone encircle the entire site. The stone cladding is laid to follow the slope of the top cap appearing to grow out of the park landscape. A weathering steel wall, made up of 376 panels, extends 61meters into the landscape, protecting the Arrivals Receiving Building. The material is perforated for translucency and is tapered as it moves away from the structures, blending into the landscape. The Arrivals Receiving Building is constructed using architectural concrete to achieve a greater workability with very dense rebars required for high pressure as well as a higher quality finish. All buildings have an overall unified appearance and the green roof complements the timber columns used at the clubhouse.

P164 Royal HaskoningDHV

Richard van den Brule is a Dutch architect with a MSc in Architecture from the TU Delft. Is a Registered Architect in the Netherlands (SBA) and a member of the Royal Institute of Dutch Architects (BNA). Has several years of (inter)national experience in a wide range of projects from designing interiors, master planning, bridges, tunnels and highways, educational buildings, railway stations, to industrial facilities. Received the Award for Best Building of the Year 2016 from the BNA for his Underground Parking in Katwijk aan Zee. From 2015 onwards, he managed the architectural department (25 people) of RHDHV in Jakarta, Indonesia and recently moved back to the Netherlands.

P62 COBE

Was founded in Copenhagen, Denmark by Danish architect, Dan Stubbergaard in 2005. Employs around 100 dedicated architects, constructing architects, landscape architects and urban planners. Has been particularly successful within the area of urban planning with a number of award-winning designs of buildings and urban spaces. Received the prestigious Golden Lion Award at the Venice Biennale in 2006, Nykredit's Architecture Prize in 2012, Dreyer's Foundation's Honorary Award in 2015 and the Eckersberg Medal in 2016.

P16 ALA Architects

Was founded in 2005 by four partners: Juho Grönholm, Antti Nousjoki, Janne Teräsvirta, and Samuli Woolston. Is today lead by Grönholm, Nousjoki, and Woolston, working with 38 architects, interior designers, students and staff members, representing 14 nationalities. Specializes in cultural buildings, terminal design and unique renovation projects. Seeks for fresh angles, flowing forms and surprising solutions on all levels of architecture. The designers challenge themselves to provide alternatives, develop prototypes and look for innovations. All three partners have around 20 years of professional experience, mostly in designing large public buildings both in Finland and abroad. In 2012, they received the prestigious Finnish State Prize for Architecture.

P122 Phil Roberts

Is a design writer based in Montreal, Canada. Also works as a design consultant for various companies in the creative industries. Has an Honours Bachelor of Arts from the University of Toronto, where he majored in architectural design and minored in Canadian studies and Spanish.

P148 JAJA Architects

Was founded in 2008 (Copenhagen) and is headed by Jakob Steen Christensen[left], Kathrin Susanna Gimmel[center] and Jan Yoshiyuki Tanaka[right], an international collective from Norway, Switzerland, Japan and Denmark. JAJA means Yes Yes! It represents their approach to architecture, which is not about a defined architectural style but rather an optimistic curiosity that explores all the possibilities that architecture can be. Their goal is to create beautiful and functional solutions which harmonize with the distinct qualities of the specific location.

P164   **OKRA Landscape Architects**
Is a design office for landscape architecture and urban planning based in Utrecht, The Netherlands. In the past 25 years, it has specialized in the development of the (urban) landscape. Its approach is to realize projects that bring people together. Is strongly committed to the development of concepts for the healthy and resilient city. From this perspective, it is crucial to work in a multidisciplinary way. Has various disciplines but likes to embrace the knowledge of experts from other disciplines to come to an integral plan.

P178   **Carrilho da Graça Arquitectos**
João Luís Carrilho da Graça graduated from the Lisbon School of Fine Arts in 1977, and began his professional practice. Was distinguished with several prestigious awards for architectural projects like Carpinteria pedestrian bridge (2012) in Covilhã, and Musealization of the Praça Nova Archaeological site at Saint George's castle (2010) in Lisbon, Santa Catarina Residences (2008). Participated in exhibition projects for Venice Architecture Biennale four times (in 2010, 2012, 2016, 2018). He is also a reputable educator, whose career mainly includes: lecturer at School of Architecture of the Technical University of Lisbon between 1977 and 1992, and a guest professor since 2014; a professor at the Universidade Autónoma de Lisboa from 2001 to 2010 and at the Universidade de Évora since from 2005 to 2015, and head of the architecture department on both institutions until 2010, and so on.

P104   **Andrew Bromberg at Aedas**
Is global design principal at the international architectural practice, Aedas. Received his Master's degree in architecture at the Southern California Institute of Architecture and University of Washington after completing his Bachler's degree in Environmental Design from the University of Colorado and Arizona State University. A keen hiker and traveler since young age, he draws inspiration from nature and his interest in how people live in different cultures to create meaningful, people-centric architectural works.

P26   **aLL Design**
Is an international collective of young and established architects and designers, each with broad artistic interests and specialisms. Was established in 2011 by Prof. Will Alsop (1947-2018), an British architect and artist. He was awarded the RIBA Stirling Prize in 2000 for Peckham Library, London and the first RIBA Worldwide Award in 2004 for Sharp Centre for Design, Toronto. His core values are innovation and expression with an emphasis on enjoyment - his practice is founded principally to 'make life better'. Was a member of architectural advisory boards for Wandsworth and Kensington & Chelsea Councils.

P16   **Esa Piironen Architects**
In 1990, Finnish architect Esa Piironen has established his own office, Esa Piironen Architects. Has completed a wide variety of projects from small-scale street furniture design to large-scale urban design, many of which are based on winning public competition entries. Always practices environmentally responsible architecture, constantly committing to humanistic principles. Esa Piironen has lectured on subjects related to urban design, architecture, environmental design, exhibition design, and environmental psychology in Finland and abroad. Was appointed as visiting professor at Guangdong University of Technology - School of Art and Design starting from 2012.

P76 AREP

Is a multidisciplinary practice in transforming the city. Founded in 1997 within SNCF (the French national rail operator) by Jean-Marie Duthilleul and Etienne Tricaud - both architects and engineers. AREP Group delivers projects on different scales going from large metropolitan areas and urban districts to individual buildings and street furniture. Its reputation is based on the ability to conceive and create multimodal stations in dense urban areas, in other words mobility-related spaces, and housing complex uses, and challenge technical and heritage issues. Building on this knowledge, AREP provides effective solutions for other types of public spaces. Brings together 750 people and more than 30 nationalities, professionals from diverse disciplines: architects, urban planners, designers, engineers, economists, architectural programming consultants and construction operations managers. Offers their expertise in all areas of city planning and construction: multimodal hubs and railway stations, public amenities, offices, hotels and housing, shopping centres and technical facilities.

P88 Zaha Hadid Architects

Zaha Hadid was born in Bagdad, Iraq in 1950. She graduated from the Architectural Association school of Architecture in 1972, and founded Zaha Hadid Architects in 1979. She is an architect who consistently pushes the boundaries of architecture and urban design. She completed her first building, the Vitra Fire Station, Germany in 1993. Her interest lies in the rigorous interface between architecture, urbanism, landscape and geology as her practice integrates natural topography and human-made systems, leading to innovation with new technologies. She taught Architectural Design at Yale University as a visiting professor and at the University of Applied Art in Vienna.

P132 Brooks+Scarpa

Has redefined the role of the architect and results in some of the most remarkable and exploratory designs today. They do so by looking, questioning and reworking the very process of design and building. Each project appears as an opportunity to rethink the way things normally get done; to redefine and cull-out latent potentials that exist in materials, form, construction and even financing to 'make the ordinary extraordinary.' Over the last ten years, has received more than 50 major design awards, notably 18 National AIA Awards, including the 2010 AIA Architecture Firm Award and 2010 AIA California Council Firm of the Year Award, five AIA Committee on the Environment 'Top Ten Green Projects' awards.

P228 Grimshaw Architects

Was founded by Sir Nicholas Grimshaw in 1980 and became a partnership in 2007. Operates worldwide with offices in Los Angeles, New York, London, Melbourne, Sydney, Kuala Lumpur, Doha, and Dubai, employing over 550 staffs. Their work is characterised by a strong conceptual legibility, innovation and a rigorous approach to detailing, all underpinned by the principles of humane, enduring and sustainable design. Is dedicated to the deepest level of involvement in the design of its buildings in order to deliver projects which meet the highest possible standards of excellence. Grimshaw was awarded the 2016 and 2018 AJ100 International Practice of the Year Award for the firm's breadth of work around the globe.

©Aleksi Valmunen

P194 Luís Pedro Silva, Arquitecto, Lda.

Is a studio based in Oporto, Portugal, founded in 2000. Has developed a gradual awareness of the collective importance of architecture, highlighting concerns of environmental and social character. Focuses on a broad perspective of space organization, coordination of a wide range of disciplines and scales. Their work mainly includes public and private buildings in Portugal and Africa, research and consultancy in several areas of intervention.
The managing partner, Luís Pedro Ferreira da Silva, completed bachelor and master degree (joint with Faculty of Engineering of the University of Porto), then completed Ph.D. at Faculty of Architecture of the University of Porto (where he teaches architecture since 2000).

P50 JKMM Architects

Was established in 1998 by four partners; Samuli Miettinen, Asmo Jaaksi, Juha Mäki-Jyllilä, Teemu Kurkela[from the left] in Finland. Operates actively in various areas and scales of architecture buildings, interiors, furniture, urban environments as well as renovations. Samuli Miettinen served as the principal architect of the Lahti Travel Centre project. Asmo Jaaksi is specialised in designing public buildings that reflect the values of humanity combined with practicality. Juha Mäki-Jyllilä is specialised in residential and office buildings. Teemu Kurkela focuses on the modernisation of healthcare architecture. Believes that buildings should be at their best even after decades of use and sustainable development is the key responsibility.

P214 PES-Architects

Was founded by Finnish professor and architect Pekka Salminen in 1968, Helsinki. Operates mainly in Finland and China but have also carried projects in the Gulf Region and Russia. Has operated in China since 2003 and opened Shanghai office in 2010. Sustainable design, in terms of ecological, economical and architectural quality, is an essential and self-evident part of its design solutions. Is constantly expanding its cooperative network of specialists in various sustainability and energy saving engineering fields. Is led by Pekka Salminen (Architect SAFA, President), Tuomas Silvennoinen (Architect SAFA, Design Director), Arttu Suomalainen (Architect SAFA, Design Director transport), and Jarkko Salminen (MSc, CEO).

P4 Richard Ingersoll

Born in California, 1949, earned a doctorate in architectural history a UC Berkeley and was a tenured associate professor at Rice University (Houston) from 1986-97. Currently teaches at Syracuse University in Florence (Italy), and the Politecnico in Milan. Served as executive editor of *Design Book Review* from 1983-1997. His recent publications include: *World Architecture, A Cross-Cultural History* (2013); *Sprawltown, Looking for the City on its Edge* (2006). Frequently writes criticism for *Arquitectura Viva, Architect, Lotus, Bauwelt* and *C3*. Regularly curates exhibitions on design at the Museo Nivola (Orani), and in 2015 provided two installations on urban farming at the MAXXI museum in Rome.

P62 Gottlieb Paludan Architects

Project and design manager, Marianne Jørgensen (1964) is a member of the Danish Association of Architects. Graduated from The Royal Danish Academy of Fine Arts, School of Architecture in 1995. Worked at DSB Arkitekter/Public Arkitekter (2005–2012), Format Arkitekter (2001–2005), KHR Arkitekter (1998–2001) and Lundgaard & Tranberg Arkitekter (1995–1998). Has special expertise in design, project and stakeholder management, technical installations, infrastructure and complex architectural projects and extensive experience with a wide range of assignments and participates fully in all project phases, including planning and design, construction management and technical supervision.

P38 Rick Joy Architects

Was founded by Rick Joy, based in Tucson, Arizona in 1993. Was born in Maine, USA in 1958 and graduated from the University of Arizona in 1990. Is the recipient of the 2002 American Academy of Arts and Letters Award in Architecture and the 2004 National Design Award from the Smithsonian Institute/Cooper-Hewitt Museum, and is a fellow of the AIA and RIBA. Senior Associate, Matt Luck earned his USGBC LEED accreditation in 2004. Has led in the design and detailing of the firm's institutional, civic, and residential projects since 2010. Natalia Hayes has been bringing experience working on institutional, cultural, commercial, religious and residential projects since 2010. Completed NCARB's Architectural Experience Program.

© 2019大连理工大学出版社

版权所有·侵权必究

图书在版编目(CIP)数据

基础设施之光：汉英对照 / 英国扎哈·哈迪德建筑师事务所等编；司炳月，高松译. -- 大连：大连理工大学出版社，2019.12
(建筑立场系列丛书)
ISBN 978-7-5685-2457-5

Ⅰ. ①基… Ⅱ. ①英… ②司… ③高… Ⅲ. ①建筑照明—照明设计—汉、英 Ⅳ. ①TU113.6

中国版本图书馆CIP数据核字(2020)第012999号

出版发行：大连理工大学出版社
　　　　　（地址：大连市软件园路80号　邮编：116023）
印　　刷：上海锦良印刷厂有限公司
幅面尺寸：225mm×300mm
印　　张：15.25
出版时间：2019年12月第1版
印刷时间：2019年12月第1次印刷
出 版 人：金英伟
统　　筹：房　磊
责任编辑：杨　丹
封面设计：王志峰
责任校对：张昕焱
书　　号：978-7-5685-2457-5
定　　价：298.00元

发　行：0411-84708842
传　真：0411-84701466
E-mail：12282980@qq.com
URL：http://dutp.dlut.edu.cn

本书如有印装质量问题，请与我社发行部联系更换。

建筑立场系列丛书01：
墙体设计
ISBN：978-7-5611-6353-5
定价：150.00元

建筑立场系列丛书02：
新公共空间与私人住宅
ISBN：978-7-5611-6354-2
定价：150.00元

建筑立场系列丛书10：
空间与场所之间
ISBN：978-7-5611-6650-5
定价：180.00元

建筑立场系列丛书11：
文化与公共建筑
ISBN：978-7-5611-6746-5
定价：160.00元

建筑立场系列丛书19：
建筑入景
ISBN：978-7-5611-7306-0
定价：228.00元

建筑立场系列丛书20：
新医疗建筑
ISBN：978-7-5611-7328-2
定价：228.00元

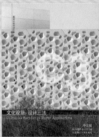
建筑立场系列丛书28：
文化设施：设计三法
ISBN：978-7-5611-7893-5
定价：228.00元

建筑立场系列丛书29：
终结的建筑
ISBN：978-7-5611-8032-7
定价：228.00元

建筑立场系列丛书37：
记忆的住居
ISBN：978-7-5611-9027-2
定价：228.00元

建筑立场系列丛书38：
场地、美学和纪念性建筑
ISBN：978-7-5611-9095-1
定价：228.00元

建筑立场系列丛书46：
重塑建筑的地域性
ISBN：978-7-5611-9638-0
定价：228.00元

建筑立场系列丛书47：
传统与现代
ISBN：978-7-5611-9723-3
定价：228.00元